Proceedings of the 30th International Geological Congress
Volume 15

Igneous Petrology

Proceedings of the 30th International Geological Congress

Proceedings of the

30th International Geological Congress

Beijing, China, 4 - 14 August 1996

VOLUME 15

Igneous Petrology

Editors:

Li Zhaonai
Institute of Geology, Chinese Academy of Geological Sciences, Beijing, China
Qi Jianzhong
Nanjing Institute of Geology and Mineral Resources, CAGS, Nanjing, China
Zhang Zhaochong
Institute of Geology, Chinese Academy of Geological Sciences, Beijing, China

CRC Press
Taylor & Francis Group
Boca Raton London New York

CRC Press is an imprint of the
Taylor & Francis Group, an **informa** business

First published 1997 by VSP BV Publishing

Published 2019 by CRC Press
Taylor & Francis Group
6000 Broken Sound Parkway NW, Suite 300
Boca Raton, FL 33487-2742

© 1997 by Taylor & Francis Group, LLC
CRC Press is an imprint of Taylor & Francis Group, an Informa business

First issued in paperback 2019

No claim to original U.S. Government works

ISBN 13: 978-0-367-44794-6 (pbk)
ISBN 13: 978-90-6764-246-0 (hbk)

**Visit the Taylor & Francis Web site at
http://www.taylorandfrancis.com**

**and the CRC Press Web site at
http://www.crcpress.com**

CONTENTS

Proc. 30ᵗʰ Int'l. Geol. Congr., Part 15, pp. 1
Li et al. (Eds)
© *VSP 1997*

Preface

The Symposium on Igneous Petrology held from 6 to 9 August 1996 during
the 30th the International Geological Congress (IGC) embraced nine topics:
volcanism in relation to tectonic setting; continental intraplate and
passive—margin magmatism; petrology, mineralogy and geochemistry of
kimberlites, lamproites, carbonatites and other alkaline rocks; magmatism
related to collisional orogenesis; thermal regimes, inclusions and tectonic
settings of granites; generation, segregation, ascent, storage and eruption of
magmas; structures and physical properties of silicate melts of magmas; the
role of water and other volatiles in magmatic processes; and new advances
in experimental petrology. A total of 362 abstracts of papers submitted to
the symposium were received. 113 papers were orally presented at nine ses-
sions of the symposium and 129 papers at poster sections.

The theme of this volume *Igneous Petrology* is intraplate (especially continental intraplate) magmatism and diversity and complexity of mechanisms
of magma formation. The contents concern such problems as difference in
magmatism of relatively stable cratons and relatively mobile continents
formed by assembly of small blocks, difference in formation conditions of
type and non—rift type magmatism of continental intraplate and their relationships, causes for the heterogeneity of continental and oceanic intraplate
magma source regions, and non—subduction mechanism for the formation
of continental high—potassium andesitic and trachyandesitic magmas.
These problems are awaited for further study.

The editors gratefully acknowledge J. Pearce, S. Kanisawa, S. Haggerty, M.
Brown, M. Wilson, L.N. Kogarko, Wang Dezi, Zhang Andi, Chen
Zhenghong, Liu Linggen, Cong Bailin, Hong Dawei, Lin Shengzhong, Mo
Xuanxue, Ma Changqian and Xia Linqi for reviewing the abstracts of the
papers submitted to the symposium.

Li Zhaonai
Qi Jianzhong
Zhang Zhaochong
Beijing, January 1997

Proc. 30thInt'l.Geol.Congr.,Part 15, pp. 3 – 12

Li Zhaonai et al.(Eds)
© VSP 1997

Cenozoic Volcanism and Lithosphere Tectonic Evolution in the Northern Tibetan Plateau, China

DENG WANMING

Institute of Geology, Chinese Academy of Sciences, Beijing, 100029. China

Abstract

There is a great volcanic belt at the northern Tibetan Plateau. It is one of the famous intracontinental volcanic belt on the Earth. On the basis of forming ages, rock association series, geochemistry and isotopic components this belt is divided into two parts: the Qiangtang sub-belt in the south and the Kunlun sub-belt in the north. They are characterized by special enriching in K2O(K2O>Na2O wt%), REE . In genesis they could be related to an initial rift zone developed in Oligocene-Miocene and an intraplate subduction zone formed in Pleistocene respectively. It is interesting that the Quaternary enriched-K2O lavas show that the higher   87Sr/86Sr(i) (0.7089-0.7105) and 206Pb/204Pb (18.711-18.834),207Pb/204Pb(15.087-15.701), 208Pb/204Pb(38.900-39.1909) but lower 143Nd/144Nd(0.5121-0.5123) than Cenozoic volcanic rocks of the East China. Perhaps these characters suggest that the primary magmas of such lavas might have been derived from partial melting of the EMII type heterogeneous enriched mantle source, but they were contaminated by continental crust during rising and eruption. The Tibetan Plateau has involved some important evolutional stages since Cenozoic. The time-space distribution and petro-geochemistry of the Cenozoic magmatisms are closely related to the lithospheric evolution.

Keywords:Volcanism, Lithosphere, Tibet, Mt.Kunlun, Qiangtang, Uplift, Cenozoic

INTRODUCTION

Relatively intense Cenozoic volcanism showed at the northern margin of the Tibetan Plateau and it's adjacent Mt. Kunlun. The regional distribution of the E-W extending volcanic belt along the Kunlun orogenic belt is a conspicuous geological feature of the Tibetan Plateau. Moreover, the eruptions of the these volcanic rocks are very important event within the Asian continent in the Cenozoic era. Because of a high topographic elevation of above 4500m, inclement climate and inaccessibility this area is among those geologically poor-studied on the Earth. Studies of the distribution, occurrence, petrology, geochemistry and isotopic chronology of the Cenozoic volcanic rocks not only are of important theoritical significance in understanding the formation and uplifting of the Tibetan Plateau, but also provide new clues to prospecting ore associated with volcanic eruptions. In this paper the petro-geochemical features of the volcanic rocks are described and it's relationship with lithosphere tectonic evolution is discussed as well.

DISTRIBUTION AND AGES OF VOLCANIC ROCKS

Studies in recent years show that this volcanic belt extends along the north Qiangtang, Mts. Kunlun and Hoh Xil from west to east, and has about 1500km in lengh, and most rock districts lie within southern Mt. Kunlun and northern Qiangtang area, only a few in the Mt. Kunlun or on it's nortern side(Fig.1).

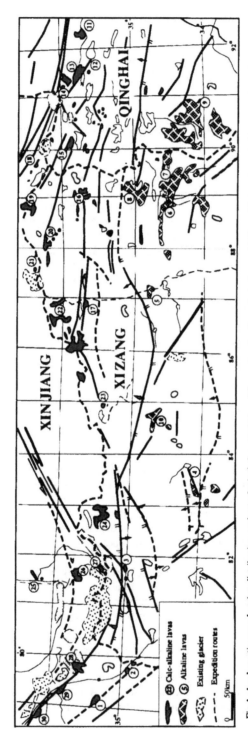

Fig.1 A schematic map showing distribution and petrological characteristics of Cenozoic volcanic rocks in the north Tibetan Plateau.(after [1-9]). Volcanic districts:
1-Tongtianqiao; 2-Hongshanhu; 3- Bangdacuo;4-Yulinshan;5-Bamaoqiongzong; 6-Duogecuoren; 7-Taipinghu; 8-Yangbohu; 9-Zhentouya,10-Xinhu; 11-Damaoshan;
12-Kekaohu(E);13-Kekaohu(NE); 14-Wuxvefeng; 15-Hctuofeng; 16-Xiangyanghu; 17-Yongpocuo; 18-Jingyuhu; 19-Xiongyintai; 20-Xuemeihu; 21-Mutztagh;
22-Qiangbaqian; 23-Yanghu; 24-Heshibeihu; 25-Pulu; 26-Ashikule; 27-Ataimupaxia;28-Quanshuigou;29-Dahongliutam; 30-Kangxiwar.

The pre-Palaeozoic and Palaeozoic strata are distributed mainly in the Mt. Kunlun area, constituting the basement of the Kunlun and the Qiangtang terranes. Mesozoic strata are of wide distribution in the Qiangtang terrane. Among them the Triassic and Jurassic systems are most developed. The Cretaceous red beds unconformably overlie older formations. The Cenozoic volcanic rocks are found overlying above-mentioned sequences and the eruptions are controlled by some regional large faults, some of which are still active.

The Cenozoic volcanic belt is composed of some provinces and includes several tens of lava districts. On the basis of study on forming times, rock associations and geochemistry this belt is divided into two parts: the Qiangtang sub-belt in the south(including west- and north Qiangtang provinces)(S-belt) and the Kunlun sub-belt(including Hoh Xil, mid Kunlun and west Kunlun provinces)(N-belt). They exhibit some similar aspects, but also many important differences each other.

We have collected many isotopic age data(Table1). It shows that the volcanic activity in the west Qiangtang province is oldest and the volcanic eruption in the west Kunlun province is youngest, other provinces are situated between above them, and there are several eruptive phases or stages in the north Qiangtang, the Hoh Xil, the mid Kunlun and the west Kunlun provinces. In general, the volcanic rocks in the Qiangtang belt are older than thos of the Kunlun belt. The former mostly was formed from Eocene to Miocene, but the latter was produced from Pliocene to Pleistocene.Moreover the age of volcanic rocks became younger from the east to the west within the Kunlun belt.

PETROLOGY,PETROCHEMISTRY AND GEOCHEMISTRY

Excepting the differences in ages, there are some obvious differences of petrology and rock association serieses in these both sub-belts. The Bamaoqiongzong district(No.5 in Fig.1) is a typical example of the S-belt. A well- shaped crater is located on the northeastern side of the main peak(Fig.4). The effusive rocks consist of a suite of ultra-alkaline lavas including augite-bearing leucitite, pseudo-leucite porphyry,leucite phonolite, noselite phonolite and trachyte, as well as syenite porphyry[1].In the TAS diagram(Fig.2a) these rocks are plotted in the alkaline area and therefore belong to alkaline series. Comparatively, volcanic rocks from the N-belt have a high content and a wider range of SiO_2. hence are concentrated mainly in the field of trachyandesite and basaltic trachyandesite, with a small proportion falling in the fields of trachyte, rhyolite,etc. All the data points of volcanics in the N-belt plot in the SHO area, is different from the High-K CA.(Fig.2b).

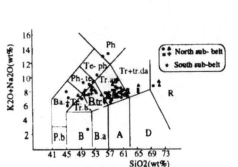

Fig 2a TAS chemical classification of volcanic rocks in the north Tibetan Plateau.(after [18])

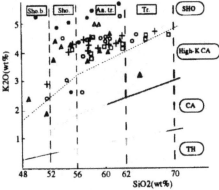

Fig.2b K2O-SiO2 plot of volcanic rocks in the north Tibetan Plateau.(after [19])

Table 1 Isotopic ages of Cenozoic volcanic rocks

Prov.	Dist.	No.	Age (Ma)	Method	Reference
QIANGTANG(W)	Tongtianqiao	1	60.0	Ar/Ar	[6]
	Bangdacuo	3	44.0	Ar/Ar	[6]
QIANGTANG(N)	Bamaoqiongzong	5	26.5	K/Ar	[10]
	Duogecuoren	6	10.6	K/Ar	[8]
	Zhentouya	9	44.66, 4.27	K/Ar	[9]
HOH XIL	Damaoshan	11	17.3, 15.39	K/Ar	[11]
	Kekaohu(E)	12	13.2, 11.7	K/Ar	[11]
	Kekaohu(NE)	13	17.6, 19.6	K/Ar	[9]
			11.7, 14.47	K/Ar	[11]
	Wuxvefeng	14	21.24, 32.9,	K/Ar	[9]
			24.55, 41.9		
	Hetoufeng	15	12.5, 18.3	K/Ar	[9]
			7.09, 12.6	K/Ar	[11]
	Xiangyanghu	16	7.49, 6.95	K/Ar	[11]
			13.3, 8.5	K/Ar	[12]
	Yongpohu	17	9.4	K/Ar	[6]
KUNLUN(M)	Jinyvhu	18	0.69	K/Ar	unpublished
	Xiongyintai	19	1.08	TL	[6]
	Mutztagh	21	4.6	Ar/Ar	[13]
	Qiangbaqian	22	14.9	K/Ar	[6]
	Yanghu	23	16.5	Ar/Ar	[6]
	Heshibeihu	24	0.067	TL	[6]
KUNLUN(W)	Pulu	25	1.1, 1.24, 1.4	K/Ar	[14]
			1.43, 1.21	K/Ar	[17]
	Ashikule	26	0.2-2.8	K/Ar	[15]
			0.074	TL	[6]
	Ataimupaxia	27	0.56	TL	[6]
	Quanshuigou	28	3.6-6.4	K/Ar	[16]
	Dahongliutan	29	0.28	TL	[6]

(TL-thermoluminescent dating)

As viewed from the chondrite-normalized REE distribution patterns(Fig.3), both sub-belts exhibit very strong enrichment in LREE, and the S-belt is more than the N-belt. In a word, the Cenozoic volcanic rocks at the northern margin of the Tibetan Plaeau are characterized by the enrichment in total alkalis , LREE and LILE.

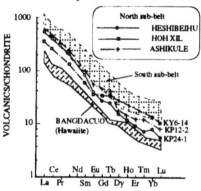

Fig.3 Chondrite-normalized REE distribution patterns of Cenozoic volcanic rocks in the Tibetan Plateau.

Fig.4 A well-preserved crater in Bamaoqiongzong volcanic district(No.5 in Fig.1) of the northern Qiangtang prov. which consists of a suite of ultrapotassic lavas i ncluding luecitites etc.

The MORB-normalized spider diagram of volcanic rocks of both sub-belts clearly display negative anomalies for the HFSE such as Nb,Ta, and Ti and for all transitional elements. Positive anomalies for Rb, Ba and Th are observed, therefore they belong to the typical post-collision geochemical environment.

ROCK-FORMING MINERALS

A common and general petrographical feature of two Cenozoic volcanic belts in northern Tibetan Plateau is having porphyritic texture, some rocks contain a few volcanic glass. Olivines, pyroxenes, hornblendes and feldspars are very important phenocrystal minerals. Forty five representative chemical compositions of rock-forming minerals have been analysed.

Olivine, partially as a phenocryst, commonly appears in some volcanic rocks of the west Kunlun province.Petrographically olivine can be divided into 2 generations: the early is mainly an euhedral phenocryst, larger in size; the latter occur as secondary phenocrysts or coexists with other microcrystals and glass in the groudmass. The content of this mineral reaches 2-5%(vol) of the rock. An olivine with Fo ranging from 71-86 is called chrysolite. Fo in olivine of the Tertiary lava is higher than that in the Quaternary lavas(averaging Fo=85 and 76, respectively), suggesting that the former is more enriched in MgO.

Where the rocks contain two kinds of phenocrysts, olivine and pyroxene, pyroxene always crystallized after olivine. The most common mineral associated with pyroxene is plagioclase. Clinopyroxenes rich in CaO widely appear in the Quaternary lavas; Orthopyroxenes poor in CaO is secondary in order. Usually, two or three generations of pyroxene can be identified(in phenocryst and matrix), and most are euhedral. There are two kinds of pyroxene coexisting with plagioclase in latite in many districts of the mid-Kunlun province. All orthopyroxenes belong to hypersthene according to pyroxene classification.Clinopyroxene is one of the most important rock-forming minerals of potassic rocks in the studied area, and usually occurs as euhedral columnal crystals. The average compositions of clinopyroxenes from both the Tertiary hawiite in the Bangdacuo and from leucitite in the Bamaoqiongzong are $En38.6Wo49.7Fs11.7$, while other clinopyroxenes in the Quaternary lavas are $En45.3Wo41.2Fs13.5$. Clinopyroxenes rich in CaO and poor in MgO in the Tertiary lava mean that the evolutional trend of magmas was toward increasing MgO but decreasing CaO. The pyroxenes rich in MgO occur in the east part of the mid Kunlun province, but the pyroxenes rich in FeO in the west. Al2O3 content in clinopyroxene has an important implication for petrogenesis in terms of experimental petrology. It is obvious that some rough estimated values of P-T conditions show fractional differentiation at an high pressure corresponding to the garnet-pyrolite interval already began in the Bangdacuo(about D=82-98km). An other magmatic pockets could be mid-pressure or even in the crust.

Plagioclase is a common phenocryst in latite and trachyandesite. It is usually subhedral or well-rounded; zoning is rare but albit-twin common. Some acid plagioclases found in quartz porphyry and rhyolitic porphyry are albite or oligoclase with An10-20. But plagioclase from most districts is andesine with $An45.6Ab45.9Or8.5$ except that plagioclase, coexisting with olivine(for example in Pulu dis.) is plotted in the labradorite area[5].K-feldspar appears only in some Quaternary volcanic districts. In latite and trachyte of the Dahongliutan a kind of alkali feldspar characterized by $Or52.6Ab45.7An1.7$ occurs as giant phenocrysts up to several centimeters long and is plotted in the sodium sanidine field.

ISOTOPIC COMPOSITIONS

The Sr, Nd and Pb isotopic compositions were analyzed in Institute of Geology,CAS and Clermont-Ferrand, France. All rocks are radiogenic in terms of Sr and unradiogenic in terms of Nd. 87Sr/86Sr(i) of Cenozoic volcanic rocks in studied area varies in a larger range from 0.704634 to 0.710540. These data can be identified to three groups: the first one is represented by the Tertiary volcanic rocks in the Bangdacuo dis. and the Tongtianqiao dis., which has lowest 87Sr/86Sr(0.704634-0.706296); The second group including some volcanic districts in the northern sub-belt has a moderate values of 87Sr/86Sr(0.707537-0.708960); And another group has an highest values of 87Sr/86Sr(0.709832-0.710540), this is a obvious feature of the youngest volcanic rocks from the Ashikule volcanic group in the west Kunlun province. In Sr-Nd spider diagram(Fig.4) these Quaternary and a few Tertiary lavas plot well below bulk earth values, and toward a crustal-type enrichment. They also plot in the same place as the Roman province high-K lavas, this is agree with the studied results of Liu et al [14] and Arnaud, et al[16]. But the Tertiary volcanic rocks (for example in the Bangdacuo and the Tongtianqiao districts) plot near the bulk earth values. This is an other difference among them. In other words, the former is characterized by higher 87Sr/86Sr and lower 143Nd/144Nd, but the latter is just opposite, their Sr and Nd isotopic components are consistent with the East Africa rift alkaline lavas. Sr and Nd isotopic compositions of above Quternary and some Tertiary volcanic rocks show that their primitive melts were not derived from partial melting of a normal sub-continental mantle source, but might be restricted to a geochemical heterogeneous enrichment mantle domains. Perhaps this enriched mantle source corresponds to the "crust-mantle mixed layer" or "crust-mantle transitional zone" discovered by geophysics in recent study[20,21]. In Fig.5 a EMII source clearly is supported by Pb, Sr and Nd isotopic components of the Quaternary lavas, it differs greatly from EMI attribution decided by the Quaternary potassic lavas produced from Wudalianchi district in the northeast China. In genesis the EMII abnormal enriched source might be related to that some oceanic sedimentary mass was brought donwwards into the mantle accretional wedge along the subduction zones during formation of some old suture zones, then a chemical mixing between the crustal materals and mantle took place near the Moho boundary.

APPROACH TO PETROGENESIS OF VOLCANIC ROCKS AND LITHOSPHERE TECTONIC EVOLUTION OF THE TIBETAN PLATEAU

Initial rift formation of the Qiangtang sub-belt

The geological evolution of the Tibetan Plateau has involved a long and complex historical process. The Neogene is an important transitional stage from microplate association one terrane after another to intense uplift. At that time, the Qiangtang area was not very high in relief. The Tibetan continental crust is characterized as being lighter and hotter than it's adjacent continental crust and thus it is easy to deformation. In fact, there is no direct connection between the eruption of Neogene volcanic rocks and the Indian plate subduction and compression on Tibet. If the formation of the Qiangtang sub-belt was directly related to this mechanism, we would expect extensive volcanic activity over much of the Tibetan Plateau for much of the past 40 Ma. There are indications that the fault-block tectonic movement took place in northern Qiangtang area due to the differences in uplifting time and scope of adjacent blocks along some weak segments of old suture zones during the Neogene. Further development of this structure would inevitably lead to the formation of an initial rift as a result of the humping of upper mantle materials. Eruptions of the Neogene volcanic rocks in the S- sub belt are probably related to the initial rift zone. It's unique petrological, geochemical and Sr-Nd-Pb isotopic characteristics show that the magma might have resulted from partial melting of the "crust-mantle mixed layer" located at near the Moho. However, this initial rift zone is of short life, and unable to full develop by the change in structural stress and the rapid rising of the Tibet since Neogene to Quaternary. Up to now, some special geomorphologic units as fault depression-basin, rift valley lakes distributing along this volcanic belt have been found.

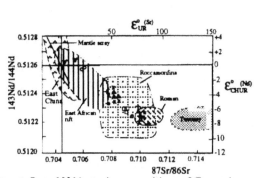

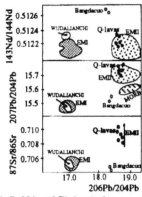

Fig. 5 Sr and Nd isotopic compositions of Cenozoic volcanic rocks from the north Tibetan Plateau. (after [6])

Fig.6 Sr,Nd and Pb isotopic components of Cenozoic volcanic rocks from the north Tibetan Plateau.

Intraplate subduction and genetic environment of the Kunlun sub-belt

Since the Late-Pliocene the whole Tibet and Mt. Kunlun were at a rapid rising stage. Some original old structures of Mt. Kunlun became active again and some large-scale strike-slip faults, thrusts and napp structures appeared. It is more important that these geological events were accompanied by the formation of the Tibetan Plateau. Therefore, the extensive Cenozoic volcanic activity is one of the forms of neotectonic movement in northern Qiangtang terrane and Mt.Kunlun.

There are evidence that the volcanic rocks in the Kunlun sub-belt are relatively close to the boundary between Mt.Kunlun and the Tarim terrane in the north in genesis. Their space-time distribution, petrological , geochemical and isotopic features can be interpreted by a model of intraplate subduction. There are indications that the neotectonic uplift of Mt. Kunlun and the subsidence of the Tarim Basin happened almost at the same time. Continued compression on the Tibet from the Indian plate met with the obstruction of the Tarim terrane from the north, resulting in a N-S shortening and thickening of the Tibetan crust. Moreover, the northern margin of the Tibet certainly obducted over the southern margin of the Tarim Basin. In other words, the southward subduction of the rigider Tarim terrane beneath the Tibetan crust,could be an important mechanism for the uplift of the Mt. Kulun and strong Quaternary volcanic eruptions in the N-sub belt. Petrological, geochemical and isotopic characteristics of these volcanic rocks demonstrate that their melts were derived mainly from fractural-compression remelting of the"crust-mantle transitional layer" owing to intraplate subduction. Even a part of the acid magmas might be derived from partial melting of the continental crust. Rapid uplifting of Mt.Kunlun took place during Late Pliocene-Early Pleistocene. But the whole and regional tectonic movements and large-scale rising of Mt.Kunlun may taken place during Late Early Pleistocene to Early Middle Pleistocene. Southward subduction of the Tarim terrane and major volcanic eruptions were most violent during this period.

There is evidence that the intraplate subduction between the northern margin of the
Tibetan Plaeau and the Tarim Basin is still under way at present. A geological survey
indicates that there is a series of overthrusts on the northern flank of Mt.Kunlun. As a
result, along these thrusts some old strata overlie the Tertiary or Quaternary deposits of
the Tarim Basin, just as the Himalayan rock series overthrusted on the Siwalik and
Ganges alluvial deposits along the MBT. According to a geophysical survey, the crust in
the southern Tarim Basin is about 40km in thickness, and is thickened to 50-70km
toward the Tibetan Plateau. In fact, there is a sudden change in crustal thickness over
quite a small distance. It implicates that the Moho discontinuity under the continental
crust dips southwards. A strong seismic zone is developed from the Hindukush to the
west through Mts. Karakorum and Kunlun to the east. The main compressional stress
axes obtained from the fault-plane solution are oriented in northeast-southwest or north-
south directions, just normal to the strike of Mt.Kunlun and volcanic belt, with a
southward dipping angle of about 60 degree [22]. There is evidence suggesting that the
Tarim Basin and Mt.Kunlun is converging recently at an average rate of about 6+4mm
per year[23].

Lithosphere evolution and it's relationship to volcanic activity and uplifting of

theTibetanPlateau since Cenozoic

The Tibet area was enclosed by the Indian, the Yangtze and the Tarim plates from south,
east and north sides. All structural compressional stress from different sides were oriented
towards the Tibet. This tectonic framework has a very important geodynamic significance
for the lithospheric evolution and uplifting of the Tibet. Therefore, a series of great
geological events has happened one after another since Cenozoic(Fig7): 1, The Yarlung
Zangbo suture zone and the Gangdies magmatic island arc have formed about 45-50Ma
ago. It marks the closing of the Neo-Tethys and finishing of the microplate accreting
history towards Eurasian continent; 2, After that, A-type subductions have taken place
along MBT and KBT(Kunlun boundary thrust) by compression from both sides of the
Indian and Tarim blocks, it results in gradual shortening and thickening of lithosphere
and slowly uplifting of the Tibetan crust. In the stage the Qiangtang volcanic sub-belt
probably formed along an initial rift zone; 3, From Late Pliocene to Quaternary,

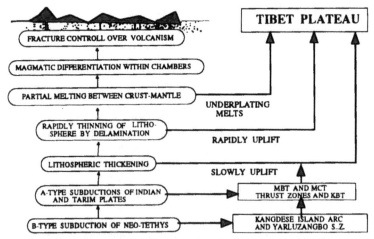

Fig.7 Tectono-volcanic activity chain showing lithospheric evolution of
      the Tibetan Plateau since Cenozoic

the thickened lithosphere mantle might be underwent a catastrophic thinning by delamination or decollement, as a result, partial melting of so-called "crust-mantle mixed layer" was caused by the upward moving of asthenosphere materials and thermal fussion of the base of lithosphere after delamination. These events necessarily led to violent uprising of the Plateau because of gravity balance and underplating of the melts; 4, Generating magmas produced from compensational partial melting should move upwards and then erupted to surface after early fractional differentiation within magma pockets along some active fractural zones. In genesis,it is most likely that the volcanic rocks in the Kunlun sub-belt may be related to this mechanism. Activity of last fault played a very important role for eruption. it will decide eruptive localities and age of the lavas. Without this condition, these melts will be preserved within crust and slowly crystallize as form as "low velocity layer" discovered by geophysics[20,21].

CONCLUSIONS

A suite of post-collisional potassic volcanic rocks widespread in north part of the Tibetan Plateau are divided into the Qiangtang and the Kunlun sub-belts respectively according to differences in petrological, geochemical characteristics and the isotopic compositions and ages. Their common features are enriching in K2O, LREE and LILE, but volcanic rocks from the Qiangtang area exhibit mostly strong short life rift attribution, while the volcanic rocks in the Kunlun sub-belt, perhaps is related closely to a southward intraplate subduction zone beneath the Mt.Kunlun from the Tarim terrane. It is supported that the volcanic melts might be derived from an EMII type enriched sources called "crust-mantle transitional layer".

The Cenozoic volcanic activities in studied area are one of displaying forms of the neotectonic movement. The Tibet was involved in a complex evolutional process since Cenozoic including a violent northward and southward A-type subductioins, the shortening and thickening of lithosphere, the catastrophic tectonic delamination of lower lithosphere, the partial melting of the crust-mantle boundary transitional layer as well as regional volcanic eruptions controlled by a series of larger scale active fractural zones.

The formation and the uplifting of Tibetan Plateau is not only of a greatest and new geological event in the Asian continent in Cenozoic, but also occupies an unique place in global tectonics. A lot of information regarding deep geology and geodynamics can be provided according to study on Cenozoic volcanism.

REFERENCES

1. Deng Wanming. A preliminary study on the petrology and petrochemistry of the Quaternary volcanic rocks of northern Tibet autonomous region, *Acta geologica Sinica*,V52, **2**, 148-162(1978)(in Chinese with English Abstract).
2. Deng Wanming. Geology, geochemistry and age of shoshonitic lavas in the Central Kunlun orogenic belt, *SCIENTIA GEOLOGICA SINICA*, **3**, 201-213 (1991) (In chinese with English Abstract).
3. Deng Wanming. Cenozoic volcanism and intraplate subduction in northern margin of the Tibetan plateau, Chinese. J. of Geoch. Vo.10, **2**, 140-152(1991).
4. Deng Wanming. Petrological and geochemical characteristics of potassic volcanic rocks in north Tibetan plateau. In 《Exploration of volcanoes and rocks in Japan, China and Antartica》 Commemorative papers for Prof. Yukio Matsumoto . 451-459 (1992).
5. Deng Wanming. Mineralogical featurtes of rock-forming minerals in Cenozoic volcanic rocks from north Tibet, China. *Scientia Geologica Sinica.* Vo.1

N3/4, 135-147(1992).

6. Deng Wanming. Study on trace element and Sr, Nd isotopic geochemistry of Cenozoic potassic volcanic in north Tibet,*Acta Petrologica Sinica*, V9, **4**, 379-387(1993) (in Chinese with English Abstract).

7. Pan Guitang, Wang Peisheng, XuYaorong, Jiao Shupei, iang Tianxiu. *Cenozoic tectonic Evolution of Qinghai-Xizang Plateau.* Geological Publishhing House. Beijing(1990)(in Chinesewith English Abstract).

8. Li Cai, Fan Heping, Xu Feng. Lithochemical Characteristics of Cenozoic Volcanic rocks in Qinghai-Xizang(Tibet) and its Structural Significance, *Geoscience*, V3, 1. 58- 69(1989)(in Chinese).

9. Zhang Yifu and Zhen Jiankang, *An Geological Introduction to Mt. Hoh Xil a nd Adjacent area.* Seismological Press. Beijing(1994)(in Chinese).

10.Deng Wanming. Cenozoic volcanic rocks in the northern Ngari district of the Tibet (Xizang) - Discussion on the concurrent intracontinental subduction, *Acta Petrologica Sinica*, 3,1-11(1989)(in Chinese with English Abstract).

11. Deng Wanming, Zhen Xilan and Yukio Matsumoto. Petrological characteristics and ages of cenozoic volcanic rocks from the Mt.Hoh Xil, Qinghai China, *Acta Petrologica et Mineralogica*(1997),(in press) (in Chinese).

12. Simon Turner, Chris Hawkesworth, J iaqi Liu, Nick Rogers, Simon Kelley and Peter van Calsteren. Timing of Tibetan uplift constrained by analysis of volcanic rocks. *Nature*, V364,50-53(1993).

13. Peter Molnar, B. Clark Burchfiel, Zhao Ziyun, Liang K'uangyi, Wang Shuji, Huang Minmin.Geologic Evolution of Northern Tibet: Results of an Expedition to Ulugh Muztagh, Science,V235,299-3059(1987).

14. Liu Congqiang et al, Sr, Nd, Ce, O isotopic and trace element geochemistry of the Kangshulak Cenozoic volcanic rocks, Yutanxian, Xinjiang, *Chinese Science B ulletin*, 23: 1803-1806(19890 (in Chinese).

15. Liu Jiaqi and Maimaiti Yiming. Distribution and ages of Ashikule volcanoes on the West kunlun Mountains, west China. *Bulletin of Glacier Research 7*, 187-190 (1989).

16. Arnaud N.O; Ph.Vidal; P.Tapponnier; Ph. Matte and Deng Wanming, The hig-K2O volcanism of northwestern Tibet: Geochemistry and tectonic implications. Earth. Plane.Sci lett. 111. 351-367(1992).

17. Liu Giaqi. Some questions on "the volcanic rocks and it's time problem in Pulu,Xinjiang" *Acta Petrologica Sinica*, **2**,95-97(1989)(in Chinese).

18. Le Maitre, R.W.(ed). A classification of igneous rocks and glossary of geological terms, Blackewell, Oxford, 193.

19. Peccerillo, A. & Taylor, S.R. Geochemistry of Eocene calcalkaline volcanic rocks from the Kastamonu area, Northern Turkey, *Contr. Miner. Petr.* **58**,63-81(1976).

20. Xiao Xuchang et al. Tectonic evolution of the lithosphere of the Himalayas, general Principle, Geological Publishing House, Beijing,(1988)(in Chinese).

21. Wu Gongjian, Xiao Xuchang and Li Tingdong. The Yadong-Golmud geoscience section on the Qinghai-Tibet Plateau,*ACTA GEOLOGICA SINICA*, V63,4, 285-296(1989)(in Chinese with English Abstract).

22. Teng Jiwen et al. The characteristics of seismic activity in Qinghai-Xizang (Tibet) Plateau and it's marginal region of China, in "Himalayan geology" (II), part of achievements in geoscientific investigation of Sino-Frence cooperation in the Himalayas in 1981. 311 - 329 Geological Publishing House, Beijing (in Chinese with English Abstract).

23. Peter Molnar, B. Clark Burchfiel, Liang K'uangyi & Zhao Ziyun. Geomorphic Evidence for acitive faulting in the Altyn Tagh and northewm Tibet and qualitative estimates of it's contribution to the convergence of Intia and Eurasia, *Geology*. 15,249-253(1987a).

Proc.30th Int'l.Geol.Congr., Part 15 pp. 13 – 21
Li et al.(Eds)

Experimental Research on Phase Transition of Al-rich Minerals in Upper Mantle of Eastern China and Its Significance

FAN QICHEN, LIU RUOXIN AND LIN ZHUORAN

Institute of Geology, State Seismological Bureau, Beijing 100029, China

XIE HONGSHEN AND ZHANG YUEMIN

Institute of Geochemistry, Chinese Academy of Sciences, Guiyang 550002, China

Abstract

The spinel/garnet lherzolite in Cenozoic basalt of eastern China and the Al-poor pyroxene bearing garnet lherzolite in ultrahigh-pressure metamorphism zone of Dabieshan-Subei-Jiaodongnan of Center-East China directly provide us with the natural samples of phase transition from spinel to garnet and from pyroxene to garnet occurring under continental upper mantle condition. We have carried out high temperature and high pressure experimental research on the above mentioned phase transitions using natural mineral and rock specimens obtained from mantle xenolith in Cenozoic basalt as the initial materials for the experiment. The experimental results provide the information about the petrographic characteristics and the change of mineral chemical composition of phase transition under different P-T conditions. We believe that there exists a phase transition zone of spinel lherzolite to garnet lherzolite with thickness of few km to about ten km at the depth of 55-70km, and we infer that the upper mantle garnet peridotite is transformed with the increase of depth from Al-rich pyroxene garnet peridotite (70-120km) to Al-poor pyroxene garnet peridotite(more than 120-150km) in continental upper mantle of eastern China

Keywords: Spinel-Garnet Phase Transition, Pyroxene-Garnet Phase Transition, Experimental Research

EXPERIMENT PROCEDURES

Study of phase transition of mantle Al-rich minerals has important significance in all these aspects as constituent of mantle, mineral association and its layering structure and tectonic deformation induced by the phase transition. The phase transition of spinel to garnet is one of the common known that Al-rich mineral phase transition occurring under upper mantle condition. The phase transition of Al-pyroxene to garnet+Al-poor pyroxen can be carried out under high pressure condition (>5.0Gpa)[1].

The experiment on phase transition of Al-rich minerals in upper mantle includes two parts. First part is on phase transition of spinel to garnet. Second part is on phase transition of pyroxene to garnet. The experimental initial material was a 200 mesh ground powder of Al-enstatite, Cr-spinel and Al-spinel chosen from the spinel lherzolite and spinel pyroxenite carried out by the Cenozoic basalt of eastern China. In order to ensure isolation of experimental specimen from the outside, preventing contact and exchange reaction between the experimental specimen and the external composition, especially Al2O3, from taking place, two ways of assembling were adopted. The first way of assembling (for specimens without water added): The specimen was put into an Al2O3 tube with tantalum foil as lining and its two ends sealed with tantalum foil; The second way of assembling (for specimens with water added): The specimen was put into a copper tube with its two end sealed and Al2O3 tube as its insulation jacket wrapped by three layers of stainless steel for electric heating. The assembled specimen was put into the center hole of a pyrophyllite cube and then into a six piston device of a YJ-3000 ton press in the laboratory of Institute of Geochemistry, CAS.

EXPERIMENTAL RESEARCH ON SPINEL-GARNET PHASE TRANSITION

PETROGRAPHIC CHARACTERISTICS OF THE EXPERIMENTAL PRODUCTS

The experimental conditions and product compositions of spinel-garnet phase transition are shown in Table1. As the time of constant temperature and pressure for every specimen was quite short (1.5-4h), we had not found any newly-grown mineral phase under optical microscope(magnification<500). However, through electron microprobe back scattering image (magnification 1000-3000), we did find newly-grown mineral in μ size (generally few μ to about ten μ). From Table 1 it can be seen that the composition of experimental products depend on starting materials and experimental conditions (P-T). Under the same temperature (T=1100°C), for the case of starting material being En+Al-Sp and under P=1.6GPa and the case of starting material being En+Cr-Sp and under P=1.8GPa, newly-grown garnet was not present in the experimental products. Meanwhile, for the above mentioned two kinds of starting materials but under P≥1.8GPa and P≥2.0GPa respectively and for the case of starting material being bulk composition of spinel lherzolite under P=2.0GPa, grown garnet and olivine were universally present in the experimental products. It can be observed between the grains of enstatite and spinel that very typical fine grown garnets (clear and without inclusions) are distributed arround spinel, while few amount of dispersed grown olivines are located at outside surrounding of garnet or near the side of enstatite (photo a,b), providing the petrographic evidence of phase transition from spinel to garnet. The content of grown garnet remarkably increases (SG9→SG6 and SG8→ SG7) with the increase of pressure. If small amount of H_2O (3%) is added into the spinel lherzolite bulk composition (SG5), the fine interstitial amphibole (about 10 μm) may be produced without finding of garnet (photo c).

Chemical composition of the experimental product

It can be seen from Table 1 that the newly-grown garnets are all pyrobe (Pyr being 64.9-76.6), while the olivines are all chrysolite (Fo being 85.6-89.6). The grown interstitial amphibole belongs to pargaside (photo c), compositionally very similar to upper mantle pargarside[2].

The mineral composition zoning of the experimental products is very remarkable. The experimental products of two kinds of starting materials have remarkably different

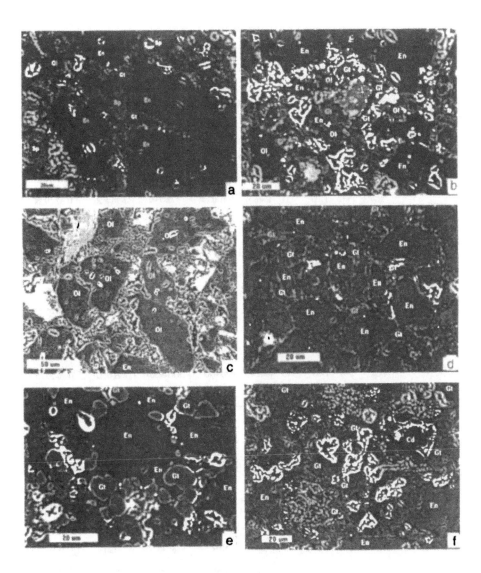

Figure 1. Back scattering image for phase transitions of spinel–garnet and pyroxene–garnet

a. SG6 Garnet(Gt) surrounded spinel(Sp), small amount of olivine(Ol) present.

b. SG7 Garnet surrounded spinel, olivine distributed outside of garnet or around enstatite(En).

c. SG5 Amphibole(Am) filling the gap between the grains of diopside(Di) and olivine.

d. PG1 Net shaped garnet surrounded enstatite.

e. PG3 Fine grain more idiomorphic garnet surrounded enstatite.

f. PG4 Mainly grown garnet with certain amount of corundum(Cd).

Fan Qicheng et al.

Table 1 Chamical compositions of experimental products for spinel-garnet phase transition

SM	No	P (GPa)	T (°C)	Time(h)	Miner.	Cr_2O_3	Al_2O_3	FeO	MgO
En:Al-Sp	SG12	1.6	1000	3	En(c-r)	0.43-0.28	5.38-6.87	6.73-8 12	32.10-30.10
4 1					Sp(c-r)	0.18-0.18	62.8-64.80	16.61-10.70	17.90-21.40
	SG9	1 8	1100	3	En(c-r)	0.41-0.29	5 68-7 99	6 70-8 20	31 52-29.38
					Sp(c-r)	0.17-0.10	64-65 30	16.33-13.92	18.40-19.70
					Gt(2)	0.24	23.30	10.29	21.85
	SG6	2	1100	2	En	0.41	5.71	6.44	31.50
					Sp	0.14	62.10	18.79	16.90
					Gt(2)	0.33	22.20	10.94	20.05
En Cr-Sp	SG10	1.8	1100	3	En(c-r)	0.38-0.72	5.82-6.86	6.51-7.13	31.38-30.59
4:1					Sp	8.06	57.39	12.53	21.28
	SG8	2	1100	3	En(c-r)	0 59-0.63	5 82-6 20	6 71-7 37	31 70-29 80
					Sp	8.62	57.80	12 16	20.87
					Gt(3)	0.95	22.00	9.64	17.80
	SG7	2.5	1150	3	En	0.56	5.65	6.12	30.90
					Sp(c-r)	10.32-19.45	56.34-48.29	12.07-10.84	20.62-19.91
					Gt(2)	2.04	21.50	7.61	20.70
Sp-lherzolite	SG4	2	1100	4	Sp(c-r)	8.76-17.61	56.98-50.54	12.16-11.17	21.62-20.13
bulk					Gt(2)	1.35	21.80	8.05	18.85
composition	SG5*	2	1100	2	Sp	8.09	55 10	11.61	21 30
					Am	0.61	20.80	8.19	15 10

SM-starting material   c-r center to rim   * add 3% H2O

composition zoning (Table 1). In the first kind of starting material (En:Al-Sp=4:1), the neighboring Cr-spinel and enstatite in the experimental product show similar Al-Cr composition zoning, with richer Al2O3 and poorer Cr2O3 at the center than at the rim of the grain, while the Mg-Fe zoning of Cr-spinel and enstatite shows different trend of variation. From the center to the rim of grain, MgO and FeO of Cr-spinel dicreases simultaneously, while MgO of enstatite decreases and FeO increases. The mineral composition zoning in the experimental products for the second kind of starting material (En:Cr-Sp=4:1) was completely different from that for the first kind. Either the neighboring mineral aluminium or enstatite shows increase of Al2O3 and slight decrease of Cr2O3 from the center to the rim of grain.

In summarizing the above, the Al-Cr zoning of minerals is the compositional zoning caused by Al, Cr ion exchange in the mineral itself, which, apart from the remarkable influence by Al/Cr ratio itself and other constituents, may be related to the degree of ion exchange between mineral grains. The Mg-Fe zoning of mineral is due to Mg, Fe ion exchange between mineral grains and is remarkably controlled by the temperature. The way of exchange between Mg and Fe ions depends on their own chemical composition. In the process of subsolid equilibrium caused by temperature decrease, the Mg-Fe composition zoning of spinel neighboring to clinopyroxene and olivine in mantle peridotite is mainly controlled by the intergrain ion exchange induced by temperature change, leading to rich Fe and poor Mg in the rim of the neighboring mineral spinel and rich Mg and poor Fe in the rim of enstatite[3]. Our experimental products result from quenching at high temperature and high pressure state equivalent to the mantle condition, leading to its difference from that mantle rock and mineral composition zoning caused by uplifting with relatively slow decrease of temperature and pressure.

Discussion

The spinel/garnet lherzolite xenolith representing the phase transition from spinel lherzolite to garnet lherzolite in the Cenozoic alkaline basalt from Minxi, Xilong, Nushan, Hebi and Hannuoba in eastern China have their the P-T equilibrium condition of T=1039-1182°C (averaged 1099°C) and P=1.89-2.05GPa (averaged 1.95GPa)[4-6].The present experimental results on spinel-garnet phase transition obtained under 1100°C and 2.0GPa using natural specimens are basically in coincidence with the P-T equilibrium condition for the spinel/garnet lherzolite from the five above mentioned areas. It is believed that the spinel lherzolite and garnet lherzolite phase transition zone in the continental upper mantle of eastern China might have been limited at the depth of 55-70km with thickness of few km to some ten km. The depth of phase transition from spinel lherzolite to garnet lherzolite in southeastern part of Australia is 55-62km[7]. It indicates the similarity between the mantles of continental lithosphere.

EXPERIMENTAL RESEARCH ON PYROXENE-GARNET PHASE TRANSITION

Petrographic characteristics of the experimental products

The experimental conditions and product compositions of pyroxene-garnet phase transition are shown in Table 2. Through electron microprobe back scattering image (magnification 1000-3000), we did find newly-grown mineral phase in μ size (generally few μ to about ten μ, see photo a, b). For the specimens with same initial composition (En+5%Al2O3) and same time of constant temperature and pressure (4h.), the content of newly-grown garnet in the experimental product increases (~10%→25%→35%) with the increase of P-T (PG→PG-2→PG-3), with its grain size increasing and its distribution as ring shape along the gap between Al-enstatite grains. When 15%Al2O3 is added into the initial composition (PG4), corundum (Cd) would appear (Photo e). For the water containing specimens (PG6 and PG7), more cavities would appear in the experimental poducts with extremely few newly-born garnets (Photo f). This is similar to the situation of water containing specimen during the experiment on spinel-garnet transition. The water has played the role of suppressing the transition from pyroxene to garnet.

Chemical composition of the experimetal products

It can be seen from Table 2 that the newly-grown phase and its chemical composition are dependent on the initial composition and the P-T condition. For the same initial composition (such as En+5%Al2O3), the pyrope molecules of garnets increase (Pyr 71.4→78.6→80.6) with the increase of P-T (PG-1→PG-2→PG-3). If small amount of water is added into the same composition as above, the pyrope molecule of the newly-grown garnets remarkably decrease. Two specimens with richest Al2O3 (PG-4 and PG-7) produce the garnet with richest pyrope molecule (Pyr 86.2-88.7). Composition zoning of Al-enstatite is very remarkable with most significant decrease of Al2O3(decrease about 1%-4%) from the center to the rim of the grain. This is completely contrary to the compostion zoning of the experimental product Al-enstatite for the transition of spinel +Al-enstatite→garnet +olivine under relatively low pressure (1.8-2.0GPa). This element migration meets the necessity of transition of Al-enstatite→garnet +Al-poor enstatite. The above mentioned transition has unlikely been carried out thoroughly in this work because the pressure was limited at 4.5-5.5GPa and the time of constant temperature and pressure was only 2-4h. However, the petrographic characteristics and the regularity of chemical change of experimental products have already indicated the ultimate trend of transition of Al-enstatite→garnet + Al-poor enstatite, which is dependent on the P-T condition, composition and lasting time of experiment.

Discussion

In recent years garnet peridotite and garnet pyroxenite of ultrahigh-pressure (3.5-5.0GPa) and relatively low temperature (<1000°C) genesis have been universally found at the ultrahigh-pressure metamorphic zone of Dabieshan-Subei-Jiaodong in Center-East China. Their pyroxenes are all Al-poor pyroxene (Al2O3 of orthopyroxene <0.2%), similar to the high pressure metamorphic ultramafic rock of Alpine [8]. Recently, Ti-clinohumite, magnesite etc. high pressure minerals (forming depth>120-150 km) have also been found at Maowu, Bixiling, Nanshanling, in Dabieshan, making the assemblage of Al-poor pyroxene-Ti-clinohumite-magnesite as a genetic indicator of ultramafic rocks of ultrahigh-pressure metamorphic zone[9]. While the pyroxene of the garnet lherzolite in Cenozoic basalt is relatively Al-rich orthopyroxene and clinopyroxene (Al2O3 5%-6% and 6%-7% respectively), with the 4-5%Al2O3 in its whole rock, which forming depth is more than 80km[4]. The pyroxenes of garnet peridotite xenolith in kimberlite regarded as from the depth of more than 100km, are rather poor Al, with Al2O3<2.5% in enstatite (most often Al2O3<1%), with Al2O3 generally less than 2.5% in diopside and with Al2O3<2% in its whole rock [4,10]. Therefore, the content of Al2O3 in pyroxene may be roughly taken as a indicator for the depth of origin of garnet peridotite and the pyroxene tende to be poor of

Al as the depth increases. In this experiment, the transition from Al-enstatite to garnet had been carried out under ultrahigh pressure condition (4.5-5.5GPa) using natural specimens. The remarkable decrease of Al2O3 along the rim of Al-enstatie indicates the trend of transition toward Al-poor enstatite. It can be expected that with the increase of pressure and time of constant temperature and pressure the percentage of garnet content in the experimental products would increase and Al-enstatite would eventually turn into Al-poor enstatite. When 15%Al2O3 was added into the initial natural enstatite, garnet and corundum appeared in the experimental product. This experimental result at the same time has supported the ultrahigh pressure genesis of red corundum garnetite coexisting with garnet peridotite [11].

Table 2 Chemical compositions of experimental products for pyroxene-garnet phase transition

No.	PG-5		PG-1		PG-2		PG-3		PG-4		
SM	En		En+5% Al2O3		En+5% Al2O3		En+5% Al2O3		En+15%Al2O3		
P(GPa)	5.0		4.5		5.0		5.5		5.0		
T(C)	1200		950		1000		1100		1100		
Time(h)	2.5		4.0		4.0		4.0		4.0		
Mineral	En(c-r)	Gt(2)	En(c-r)	Gt(2)	En(c-r)	Gt(2)	En(c-r)	Gt(2)	En	Gt	Cd(2)
SiO2	53.50-56.50	42.40	54.30-54.83	41.49	54.64-57.61	42.71	54.10-56.60	42.80	54.30	43.20	0.19
Cr2O3	0.49-0.22	1.16	0.38-0.30	0.64	0.35-0.33	0.44	0.4-0.26	0.85	0.40	0.45	0.06
Al2O3	5.72-2.13	21.82	5.79-4.40	24.03	5.51-1.83	22.89	5.80-2.49	22.95	5.55	23.20	97.14
FeO	6.74-6.29	7.99	6.29-5.59	6.99	6.47-5.18	7.56	6.54-5.36	6.92	6.30	6.48	0.45
MgO	31.5-33.2	21.40	31.9-32.75	20.85	32.03-35.06	22.26	31.60-34.30	23.10	31.60	24.90	0.18
CaO	1.11-0.90	4.18	1.09-0.78	5.47	0.97-0.25	3.33	0.93-0.55	2.90	1.11	1.28	0.00
En	89.9-90.4		90.0-91.3		89.8-92.3		89.6-91.9		89 9		
Pyr		76.4		71.4		78.6		80.6		86.2	

SM starting material   c-r center to rim   ( ) number of analysis

COCLUSIONS

In this study, experiment using starting materials of different association of natural minerals from upper mantle and under different conditions of temperature and pressure was conducted and the obtained results were applied to the explanation of phase transition in upper mantle of eastern China, leading to the following preliminary results:

1.A spinel lherzolite-garnet lherzolite phase transition zone with thickness of few km to some ten km at the depth of 55-70km in continental upper mantle of eastern China has been suggested. The upper mantle garnet peridotite is transformed with the increase of depth from Al-rich pyroxene garnet peridotite (70-120km) to Al-poor pyroxene garnet peridotite(more than 120-150km) in continental upper mantle of eastern China.

2. Pargasite was present in the experiment product with small amount of water added into the starting material but without garnet, indicating that water promote metasomatism while suppressing phase transition.

3. The Mg-Fe composition zoning in spinel (including Cr-Sp and Al-Sp) and enstatite is the result of Mg, Fe ion exchange between mineral grains, remarkably controlled by temperature, and the way of Mg, Fe ion exchange depends on their own chemical composition. The Al-Cr zoning in minerals is due to the ion exchange of Al, Cr themself, remarkably influenced by Al/Cr ratio of themself and other composition.

Some reseachers using synthetic oxides in the $CaO-MgO-Al_2O_3-SiO_2$ (CMAM) system have experimentally determined the P-T condition ($\sim 1100°C$ and 1.6-1.9GPa) of the spinel lherzolite and garnet lherzolite phase transition and the influence on the transition by adding Cr_2O_3 and a minor other constituent to the system[12-15]. Brey et al. (1990) conducted experiment on synthesizing primary lherzolite using natural specimens under the condition of 2.0GPa, 1100°C, 1%H_2O added, 72h; 2.1GPa, 1000°C, 5%H_2O added, 168.5h; and 2.0GPa, 1200°C, 13.75h; but garnet was not present in all cases[16]. However, in the present experiment, phase transition was realized using natural mantle specimens of different constituent under the conditions of relatively low P-T (T=1100°C, P=1.8-2.0GPa) and very short constant temperature and pressure (1.5-4.0h). Therefore, considering the varieties of upper mantle constituent and long geological evolution history, the real condition for upper mantle phase transition needs to be further studied.

Acknowledgements

This work were supported by the National Natural Science Foundation of China(NNSFC).

REFERENCES

1. A. E. Ringwood . The pyroxene-garnet transformation in the earth's mantle, *Earth Planet. Sci. Lett.* **2**, 255-263(1967).
2. J.F.G. Willkinson and R.W. Le Maitre. Upper mantle amphiboles and micas and TiO_2, K_2O, and P_2O_5 abundances and $100Mg/(Mg+Fe^{2+})$ ratios of common basalts and andesites: Implications for modal mantle metasomatism and undepleted mantle compositions, *J. Petrol.* **28**, 37-73(1987).
3. K. Ozawa. Olivine-spinel geospeedometry: Analysis of diffusion-controlled $Mg-Fe^{2+}$ exchange, *Geochim. Cosmochim. Acta.* **48**, 2597-2611(1984).
4. Q.C. Fan and P.R. Hooper. The mineral chemistry of ultramafic xenoliths of eastern China: Implications for upper mantle composition and the paleogeotherms, *J. Petrol.* **30**,1117-1158(1989).
5. Q.C. Fan and R.X. Liu. Study on phase transition of multiple spinel/garnet peridotite in upper mantle beneath eastern China. In: *Proccedings of upper mantle characteristics and dynamics of China.* Committee on Mantle Mineralogy, Petrology and Geochemistry, Chinese

Society for Mineralogy, Petrology and Geochemistry(ed), pp.72-82. Seismology Press, (1990).

6. Q.C. Fan and R.X. Liu. Spinel/garnet lherzolite in Late Tertiary limburgitic pipe from Hebi, Henan province and its genesis, In: *New devolopment in Mineralogy*, Petrology and Geochemistry Study in China. Z.Y. Ouyang(ed), Lanzhou University Press, 141-142(1994).

7. W.L. Griffin and S.Y. O'Reilly. Is the continetal moho the crustmantle boundary? *Geology*, 15:241-244(1987).

8. D. A. Carswell. The metamorphic evolution of Mg-Cr type Norwegian garnet peridotites, *Lithos*, 19:279-2297(1986).

9. Q. C. Fan, R. X. Liu and Q. Zhang. et al. Petrology and high-pressure mineral assemblage of mafic-ultramafic rocks of ultrahigh-pressure metamorphic zone in Dabieshan, *SCIENCE IN CHINA (Series D)*, **39:3**,329-336(1996).

10. J. B. Dawson. Kimberlites and Their Xenoliths, *New York, Springer-Verlag Berlin Heidelberg*(1980).

11. Q.C. Fan, R.X. Liu and B. L. Ma et al. The corundum garnetite of Zhimafang, Donghai district, *ACTA PETROLOGICA SINICA(in chinese)*. **8:3,** 291-295(1992).

12. I.D. MacGregor. The effect of CaO, Cr_2O_3, Fe_2O_3 and Al_2O_3 on the stability of spinel and garnet peridotites, *Phys. Earth Planet. Interiors.* **3,**372-377(1970).

13. H.St.C. O'Neill. The transition between spinel lherzolite and garnet lherzolite, and its use as a geobarometer, *Contrib. Miner. Petrol.* **77,**185-194(1981)

14. T. Gasparik. Two-pyroxene thermobarometry with new experimental data in the system $CaO-MgO-Al_2O_3$ -SiO_2, *Contrib. Miner. Petrol.* **87,**87-97(1984).

15. S.A. Carroll Webb. and B.J. Wood. Spinel-pyroxene-garnet relationship and their dependence on Cr/Al ratio, *Contrib. Miner. Petrol.* **92,**471-480(1986).

16. G.P. Brey. T. Kohler and Nickel, K.G. Geothermobarometry in four-phase lherzolites I. Experimental results from 10 to 60 kb. J. Petrol. **31,**1313-1352(1990).

*Proc. 30*th *Int'l. Geol. Congr.* · *Part* 15 · pp. 23–37
Li *et al.* (Eds)
ⓒ VSP 1997

Twin Magma Source Model of Bimodal Volcanics in Two Continent–Margin Rifts, SE China: Constraints on Elemental and Nd–, Sr– and Pb–Isotopic Studies

FANG ZHONG , XIA BANG-DONG , TAO XIAN-CONG[1],
LI HUI-MIN , ZHANG GEN-DE and LI HUI-MIN

①*Department of Earth Sciences, Nanjing University, Nanjing 210093, China*
②*Tianjin Institute of Geology and Mineral Resources, Tianjin 300170, China*

Abstract

Some elemental and U-Pb. Sm-Nd and Rb-Sr isotopic results are presented for Palaeozoic and Mesozoic bimodal volcanic suites from Hainan Island. S. China and low reaches of Yangtze River. E. China, respectively. The two bimodal suites were continental rifting products at various periods. Tholeiites of Hainan Isl. have characteristics of typical depleted mantle basalts; but the mafic volcanics from Lower Yangtze region are relatively enriched in LIL and LREE. owing to marked contamination of Proterozoic crust rocks. Felsic volcanics of the two bimodal suites developed from anatectic melting of continental crust rocks. So that. twin magma source model for the bimodal volcanic suites are suggested in the study.

Key words: Bimodal volcanics. Continantal rifting. U-Pb geochronology. Nd-Pb-Sr-isotopes

INTRODUCTION

Continental rift magmatism has been focused by petrologists, geochemists and other geologists for two decades. It is capable of providing insight into processes involving continent breakup and formation of new intra- or inter- continental basins, as well as allowing access to the chemical and isotopic signature and structure of subcontinental mantle[1,2,3]. These studies have led to the development of some hypotheses for the driving mechanism of rifting, such as they could be caused by either differential stresses in the lithosphere or asthenosphere mantle plume upwelling. The studies also indicate involvement of both enriched and depleted mantle reservoirs throughout the evolution history of the rifts with chemically and isotopically more depleted sources becoming increasingly predominant and crustal contamination less significant with time[4,5,6,7,8].

In this study, we present two sets of bimodal mafic and felsic volcanic rocks, which belong to typical volcanic associations in rifts, are located in Hainan Island, S. China, and low reaches of Yangtze River, E. China, respectively. Hainan Isl. is southern margin part of Cathaysian Plate (i. e. South China Plate); the lower reaches of Yangtze River is a depression region in eastern Lower Yangtze Plate. From Devonian starting and till Triassic, extensional conditions prevailed in the marginal parts of the two plates, in where some volcanic rocks of rifting period had developed[9,10,11,12]. The two bimodal volcanic suites are respec

-tively detailed as following.

BIMODAL VOLCANIC ROCKS OF HAINAN ISLAND

Geological Setting

Hainan Island is located off the shore of the South China Sea (Lat. 18°-21°N, Lon. 108. 5°-110°E), it is a continental-type island. Covering more than 37% of the whole island area are granitoids, including mainly porphyroblastic granites yielding Rb-Sr whole-rock isochron ages from 231. 6Ma to 320Ma, and some Mesozoic granites yielding Rb-Sr whole-rock isochron ages from 235. 1Ma to 87. 5Ma[13,14,15]. The rest of the area is occupied by supracrustal sequences as well as Cenozoic basaltic volcanics. Pre-Mesozoic stratigraphic sequences occurring in the middle band of the island consist of phyllites, slates, gneisses and minor migmatitic granites. Widespread in the late Palaeozoic formations are several layers of fluvial intermontane conglomerates whose distriibution is controlled by rift faults. Lithological and lithochemical studies indicate that the detrital rocks were formed in a tectonic setting of continental rifting. The evolution of the rifting terminated at the stage of transition from intra-continental rift to inter-continental one; the rift basin was a bay opening westward to ocean[10,17,18].

The bimodal meta-volcanics were found to occur in the western part of the island, which form an E-W trending belt extending discontinuously about 80km along Changjiang-Anding Fault. They are composed dominantly of interlayered tholeiites and quartz rhyolites, which are intimately interbedded with phyllites and slates, and form part of Shilu Group. The western sector of the bimodal volcanic rock belt is dominated by basalts with few rhyolites, earlier have a maximum single-layer thickness of over 140m; but the eastern sector of the belt is composed of rhyolites, whose maximum thickness for a single layer exceeds 20m, and some basalts; and they occur as alternations with metamorphic sedimentary rocks. These rocks contain marine of fossils such as brachiopods, bryozoans and crinoidal stems[10]. The fossils suggest this volcano-sedimentary sequence deposited during Carboniferous, coincident with Rb-Sr whole rock isochron ages in the range from 330Ma to 311Ma yielded by the meta-sediments[16].

Elemental Geochemistry of the Bimodal Volcanic Rocks

In the western sector of the bimodal volcanic rock belt (i. e. Junying and Shilu areas), the mafic rock is represented by spilites; in the middle and eastern segments of the belt are distributed by actinolitites. As shown in $FeO^* - (K_2O + Na_2O) - MgO$ diagram, the above-mentioned metamorphic basic volcanics fall into the field of tholeiite. In $TiO_2 - K_2O - P_2O_5$ diagram the majority of samples belong to oceanic tholeiites. The abundance patterns of their incompatible elements can be divided into two types: (1) the spilites at Junying, the western end of the belt (sample No. 1 in Fig. 1) possess the characteristics of N-MORB, such as low elemental abundances, gentle pattern of curve and positive Nb anomalies. (2) the tholeiites in the middle segment of the belt (Sample No. 5 and 6 in Fig. 1) and those in the eastern segment of the belt (Sample No. 10 in Fig. 1) possess the abundance patterns of E-MORB and IRT (initial rift tholeiites). However, all the tholeiitic samples from the eastern part of the belt fall into the area of within-plate basalt (WPB) on the Zr-Nb-Y and Zr-Zr/y discrimination diagrams from Meschede (1986) and Parce and Norry (1979)[19,20].

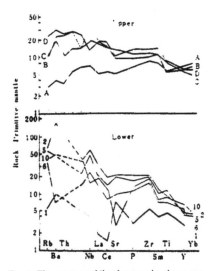

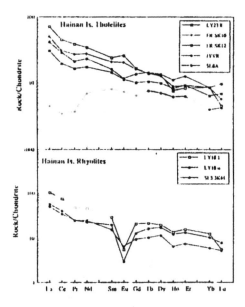

Fig. 1 The magmatophile element abundance pat terns in the Carboniferous basalts of Hainan Island. Upper: The patterns in different geotectonic set tings. A. N-MORB; B. E MORB; C. IRT; D. CT.
Lower: The magmatophile element abundance pat terns in the Carboniferous basic end-member vol canic rocks. 1. JYVR; 2. SL 6 A; 5. FR SK 10; 6. FR-SK-12; 10. LY-23 8.

Fig. 2 Chondrite-normalized REE patterns for some representative samples of the bimodal volcanic rocks from Hainan Island.

Two types of REE patterns can be distinguished for the mafic end-member of the bimodal volcanic rocks (Fig. 2): (1) Junying spilite (Sample No. 1 (JYVR)) are similar to N-MORB, characterized by slight LREE depletion, gently inclined curve, low total amount of ΣREE and positive δEu anomalies; and (2) other tholeiites like E-MORB, are characterized by LREE enrichment, rightward-inclined curves, relatively high total amount of ΣREE and positive δEu anomalies[21]. Sample No. 2 (SL-6-A) and 9(FR-SK-17) show an abnormal REE distribution patterns as a result that they were affected by later granite intrusion. However, rhyolites, the felsic end-member of the bimodal volcanic rocks, enrich in K_2O and Rb, but strongly depleted in CaO, Na_2O and Sr. All the three representative samples show moderately enriched LREE with large negative δEu anomalies(Fig. 2), which resemble those of the late Palaeozoic granite generated from remelted crust source. [14,15,16,22]

Nd- and Sr-Isotopic Compositions of the Bimodal Suites

Sm-Nd and Rb-Sr isotopic data are listed in Table 1. ϵNd(T) values are calculated using T=330Ma, this is Rb-Sr whole rock isochron age obtained from the intralayered metasediments[16]. As mentioned before, some fossil evidence suggests that this age of 330Ma should closely approximate the true depositional age of the volcano-sedimentary sequence. The spilite (Sample No. 1 (JYVR)) from western

Table 1. Sm Nd and Rb Sr isotopic analyses of the Carboniferous bimodal volcanic rocks in Hainan Island

Item	(1) JYVR	(2) SL 6A	(3) FRSK-6	(5) FRSK 10	(6) FRSK-12	(8) FRSK-14	(9) FRSK-17	(10) LY-23 8	(11) SL 5 3Km	(12) LY-18 1	(13) LY-18 6
	Spilite	Spilite		Tholeiite				Tholeiite	Rhyolite	Porphyritic quartz rhyolite	
Sm (ppm)	2.29	6.09		6.25	4.30		8.50	7.61	4.33	8.79	4.73
Nd (ppm)	5.85	27.89		25.32	17.00		42.76	35.12	20.88	41.09	22.32
$^{147}Sm/^{144}Nd$	0.23648	0.13208		0.14927	0.15308		0.12017	0.13205	0.12528	0.1297	0.1281
$^{143}Nd/^{144}Nd$	0.513086	0.512236		0.512935	0.512906		0.512318	0.512727	0.512139	0.512154	0.512158
	±11	±7		±12	±7		±6	±6	±14	±7	±7
$\varepsilon Nd(0)$	8.50	-8.07		5.56	5.00		-6.47	1.51	-9.97	-9.68	-9.61
	±0.22	±0.13		±0.23	±0.23		±0.12	±0.12	±0.27	±0.13	±0.13
$\varepsilon Nd(330Ma)$	6.82	-5.35		7.56	6.84		-3.25	4.23	-6.96	-6.86	-6.72
T_{DM} (Ma)		1518		460	545		1228	716	1562	1610	1578
Rb (ppm)	183.20		34.74	47.24	29.02	14.94	113.0	35.61	536	241	334
Sr (ppm)			142.61	458.63	218.66	308.35	205.30	489.07	19	12	28
$^{87}Rb/^{86}Sr$	0.07714		0.22731	0.29815	0.38439	0.14026	1.59713	0.210663	83.80	58.41	37.06
$^{87}Sr/^{86}Sr$	0.709806		0.71201	0.709804	0.716660	0.71234	0.73561	0.70526	0.97993	0.72222	0.85751
	±12		±4	±7	±19	±11	±56	±54	±2	±1	±2

Note: The tholeiites listed in the table have been converted to actinolite or actinolitic porphyry as a result of regional metamorphism and alteration; sample Nos. 2 and 11 collected from the vicinity of the major orebody of the shilu iron deposit were strongly altered; in the following are the constants and expressions for calculation:

The standard ratio of $^{146}Nd/^{144}Nd$ is 0.7219;

$\varepsilon Nd(T) = [[(^{143}Nd/^{144}Nd)_{sample} / (^{143}Nd/^{144}Nd)_{chur} - 1] \times 10^4$, where $(^{143}Nd/^{144}Nd)_{chur} = 0.51265$; $(^{147}Sm/^{144}Nd)_{chur} = 0.1967$;

$$T_{DM} = -\frac{1}{\lambda} \ln \left\{ 1 + \left[\frac{(^{143}Nd/^{144}Nd)_{sample} - (^{143}Nd/^{144}Nd)_{DM}}{(^{147}Sm/^{144}Nd)_{sample} - (^{147}Sm/^{144}Nd)_{DM}} \right] \right\};$$

where $(^{143}Nd/^{144}Nd)_{DM} = 0.513163$. $(^{147}Sm/^{144}Nd)_{DM} = 0.225$. $\lambda = 6.54 \times 10^{-12}/a$.

segment of the belt has the highest values of $^{147}Sm/^{144}Nd$ and $^{143}Nd/^{144}Nd$ ratios and corresponding higher $\epsilon Nd(T)$ value of 6. 82; and the tholeiite samples from middle part of the belt have intermediate values of $^{147}Sm/^{144}Nd$ and

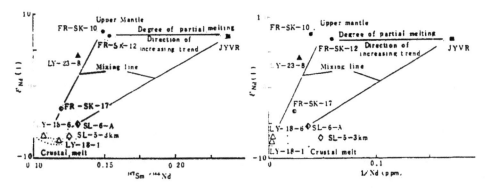

Fig. 3 Diagrams showing the Nd isotopic evolution of the Carboniferous bimodal volcanic rocks in hainan Island.

$^{143}Nd/^{144}Nd$ ratios (Table 1 and Fig. 3). But among all normal samples of the mafic volcanics, the sample (No. 10(LY-23-8)) from eastern segment of the volcanic belt has lower values of $^{147}Sm/^{144}Nd$ and $^{143}Nd/^{144}Nd$ ratios, and corresponding lower $\epsilon Nd(T)$ value of 4. 23. Owing to the alteration effect from later granite intrusion, two tholeiites (Sample No. 2(SL-6-A) and 9 (FR-SK-17) occupy the lowest values of $^{147}Sm/^{144}Nd$ and $^{143}Nd/^{144}Nd$ ratios and $\epsilon Nd(T)$ values in the mafic rocks. The two samples fall in the lower left on Fig. 3, near the field of crustal melt. Except the two samples, all normal mafic end-members of the bimodal suite indicate their derivation from the depleted upper mamtle. On account that the Junying spilite (Sample No. 1 (JYVR)) is characterized by minimum ΣREE, slight depletion of LREE, higher Zr/Nb and Y/Nb ratios, lower La/Nb, Ti/V, and Ti/Yb ratios, and isotopically by relatively higher $^{147}Sm/^{144}Nd$ and
$^{87}Sr/^{86}Sr$ values, it is different from the tholeiite samples from middle and eastern part of the volcanic belt, reflecting a higher degree of partial melting of the original upper mantle[23].

A mixed isochron was established on the basis of Rb-Sr isotope data from the basic volcanic rocks (Table 1), giving an initial $^{87}Sr/^{86}Sr$ ratio of 0. 7063, slightly higher than that of normal basalts from depleted mantle. This implies that the basic volcanic rock experienced Sr isotopic exchange during the subsequence regional metamorphism.

The acid end-member rocks (i. e. rhyolite and porphyritic quartz rhyolite) of the bimodal volcanic suites are distributed mainly in the eastern segment of the volcanic belt besides some in Shilu Iron Ore field of western part of the belt. The rhyolites yield limited ranges for $^{147}Sm/^{144}Nd$ and $^{143}Nd/^{144}Nd$ ratios and the markedly negative ϵNd (T) values from $-6. 72$ to $-6. 96$ (Table1; Fig. 3). Their Nd model ages (1. 56Ga-1. 61Ga) are similar to the U-Pb ages of rounded detritic zircons in Shilu Group phyllites (1. 35Ga)[16] and the Sm-Nd whole rock isochron age of Baoban Group migmatic gneisses (1. 70Ga)[24]. Nevertheless, the features of Nd and Sr isotopic components are full concordant with those of Hercynian-Indosinian granites in Hainan Island[25], so that the acid igneous rocks were the products from crust-source magma under high heat flow conditions during the different stages of rifting.

BIMODAL VOLCANICS IN LOW REACHES OF YANGTZE RIVER

Geological Setting

Early Triassic bimodal subvolcanics were first found in Later Permian coal-seam of Longtan Formation in Jiangsu Province, E. China by Fang et al. (1994)[28]. The bimodal suites are distributed in widespread eastern part of low reaches of Yangtze River (Lat. 30. 9°N-32°N, Lon. 118. 4°E-120°E). Tectonic setting of the study area belongs to a part of fault-depression region in the Lower Yangtze Plate. There are a few layers of volcanic ash deposits which are recently found in various marine sedimentary formation from Devonian to Triassic periods. For example, few layers of altered tuffaceous shales in middle Carboniferous limestone and in chert formation of early Permian Gufeng Formation, and the bimodal subvolcanic rocks in Later Permian Longtan Formation etc. [9,11,12,26,27,28]. So that, extensional environment in Yangtze Plate started from early Hercynian, and resulted in Triassic. But the extensional processes were strengthened by Permian time.

Elemental Geochemistry of the Bimodal Subvolcanics

Between the mafic and felsic subvolcanics, there are a hack of silica contents from 54. 4% to 64. 4%. In AFM diagram, there are identified in two kinds of basaltic

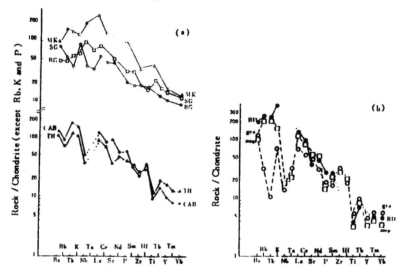

Fig. 4 (a). Chondrite-normalized spiderdiagram showing comparison between basalts of the bimodal suites from Lower Yangtze region and those from diverse rift settings.
TH-tholoiitic basalt and CAB-calci alkali basalt from the bimodal suites. MK-basanite from eastern Kenya Rift. SG-alkali basalt from South Gregory Rift. RG-transitional basalt from Rio Grande Rift (after Wilson. 1989)[29]. (b). Chondrite-normalized spiderdiagram showing comparison between rhyodacite from the bimodal suites and metamorphic crust rocks. RD-rhyodacite from the bimodal suires. gra-granulite. amp-amphhibolite (after Weaver and Tarney. 1981)[10].

rocks, e. i. tholeiitic and calc-alkalic basaltic rocks. The tholeiitic rocks are distributed at centre-southern study area (Yixing area); calc-alkalic basalts are in northern district (Nanjing and Zhenjiang areas) and in sauthern district (Guangde area) of the study region. All of the mafic volcanics belong to continental basalts as shown in 2Nb-Zr/4-Y diagram and in Zr/Y-Zr diagram[19,20]. The compositions and chondrite-normalized patterns of incompatible elements of these mafic rocks are like those of typical continental rift basalts from South Gregory rift and Rio Grande rift (Fig. 4a)[29]. On other side, The patterns of incompatible elements from the rhyodacites seem to be those of old metamorphic crust rocks, such as granulite and amphibolite (Fig. 4b)[30]. The later facts suggest that the acid end-member of the bimodal volcanic suites could be developed from ancient crust rocks remelted.

Compositions and condrite-normalized REE patterns of the bimodal suites in Lower Yangtze region show that the basaltic rocks are LREE enriched with La Contents of 50-200times chondrite, La/YbN = 1. 5 ~ 3. 6, and have negative Eu anomalies (Eu/Eu˙ = 0. 76 ~ 0. 89). The REE contents of acid volcanics are similar to average compositions of continental crust rocks, that was determined from post-Archean shales by Taylor and Mc Lennan (1985)[31].

U'-Pb Geochronology of Single-Grain Zircons

There are three kinds of zircons have been found from the bimodal volcanics of Lower Yangtze region by Fang et al. (1994)[28]. (1)Contemporaneous zircons are characterized by light-yellow colour and excellent crystal forms; the kind of zircons are main component of zircon group in a basaltic andesite. (2) xenocrystal zircons are characterized by light-purple colour, rounded form or various crystal forms; the kind of zircons are major elements of zircons from a rhyodacite sample. (3) regenerated zircons are not any colour and very lank forms. U-Pb ages of 7 single-grain zircons from CB-1 basaltic andesite, and U-Pb ages of 6 single-grain zircons from XLS-1 rhyodacite are presented in Table 2. Owing to ^{238}U and ^{206}Pb in zircons are the highest abundance isotopes, so that, we could accurately measure ^{206}Pb/^{238}U ratios to calculate ages of zircons. In some single-grain ancient xenocrystic zircons, the higher abundances of radiogenic ^{207}Pb could be also precisionally analyzed, therefore, their ^{207}Pb/^{235}U ages and ^{207}Pb/^{206}Pb ages are concordant with ^{206}Pb/^{238}U ages (Table 2)

The ^{206}Pb/^{238}U ages of contemporaneous zircons from CB-1 basaltic andesite are 231. 2±2. 5Ma; the ^{206}Pb/^{238}U ages of xenocrystic zircons from CB-1 sample are 2281Ma and 746Ma, respectively; and the ^{206}Pb/^{238}U ages of regenerated zircons from the sample are about 128. 6Ma and 156. 5Ma, respectively. In XLS-1 rhyodacite, there are two kinds of zircons, i. e. contemporaneous and xenocrystic zircons. An average ^{206}Pb/^{238}U age of two contemporaneous single-grain zircons from XLS-1 rhyodacite is 234. 5±5. 7Ma; and ^{206}Pb/^{238}U ages of four xenocrystic single-grain zircons in the sample are 1949Ma, 1792Ma, 661Ma and 542Ma, respectively(Table 2).

Almost same ages of contemporaneaus zirons from both basaltic andesite and rhy-

Table 2. U—Pb isotopic results of single—grain zircons from the bimodal volcanics in Lower Yangtze region

Sample Name	Genetic type of Zircon	Weight of Zircon(μg)	U (μg/g)	Pb (μg/g)	Common Pb(ng)	Atomic ratios [1]					Apparent ages(Ma) [2]		
						$\frac{206Pb}{204Pb}$	$\frac{208Pb}{206Pb}$	$\frac{206Pb}{238U}$	$\frac{207Pb}{235U}$	$\frac{207Pb}{206Pb}$	$\frac{207Pb}{206Pb}$	$\frac{208Pb}{238U}$	$\frac{207Pb}{235U}$	$\frac{207Pb}{206Pb}$
CB—1 basaltic andesite													
1	Contemporaneous Zircon	5	595	21	0.00013	25993	0.0657	0.03659 (87)	0.2935 (1115)	0.05816 (2036)	231.7	261.3	536
2		5	300	28	0.03	71	0.3418	0.03668 (185)	0.7633 (2336)	0.1502 (386)	233.4	575.9	2346
3		10	389	22	0.041	141	0.1290	0.03648 (47)	0.4415 (574)	0.0877 (1036)	231.0	371.3	1378
4	Xenocrystal Zircon	5	217	34	0.019	240	0.09265	0.1227 (37)	1.031 (553)	0.06093 (3089)	746	719	637
5		1	288	199	0.029	143	0.2176	0.4245 (310)	9.5300 (3469)	0.1628 (563)	2281	2390	2485
6	Metabolic Zircon	5	1874	41	0.0076	836	0.1138	0.02015 (12)	0.1406 (129)	0.05061 (483)	128.6	133.6	223.3
7		5	937	39	0.044	114	0.1367	0.02457 (47)	0.2040 (579)	0.06020 (1600)	156.5	188.5	661
XLS—1 rhyodacite													
1	Contemporaneous Zircon	5	83	4	0.0009	519	0.1851	0.03752 (237)	0.6224 (1643)	0.1203 (272)	237.4	491.4	1961
2		5	75	11	0.012	53	1.113	0.03654 (399)	2.057 (560)	0.4078 (57)	231.6	1135	3936
3	Xenocrystal zircon	5	119	18	0.017	120	0.1556	0.08774 (408)	1.111 (511)	0.09187 (3813)	542	759	1465
4		10	138	22	0.034	169	0.1829	0.1079 (23)	1.710 (347)	0.1149 (219)	661	1012	1878
5		10	442	148	0.009	4825	0.08888	0.3204 (12)	4.961 (42)	0.1123 (8)	1792	1813	1837
6		5	406	155	0.0055	3856	0.1123	0.3531 (15)	6.121 (89)	0.1257 (16)	1949	1993	2039

① The $^{206}Pb/^{204}Pb$ ratios have been corrected by experimental blank and isotopic dilution. The experimental blank is Pb = 0.030ng. and U = 0.002ng. The numbers of brackets are absolute error (2σ).

②Age: $t(^{206}Pb/^{238}U) = \frac{1}{\lambda^{238}U} \ln(\frac{^{206}Pb}{^{238}U}-1)$; $\frac{^{207}Pb}{^{206}Pb} = \frac{^{238}U}{^{235}U} \times \frac{e^{\lambda^{235}t}-1}{e^{\lambda^{238}t}-1}$.

$\lambda^{-38}U = 1.55125 \times 10^{-10}y-1$; $\lambda^{235}U = 9.84850 \quad 10^{-1} y-1$

odacite account for the isochron character of the bimodal volcanics. All xenocrystic zircons of the two samples are arranged in a row on the $^{206}Pb/^{238}U$ vs. Pb/U (wt%) relations, that indicates unvarying relationship of radiogenetic reduction-accumulation between U and Pb isotopes in xenocrystic zircons during their geological periods. A few xenocrystic zircons in the basaltic rocks are the strong evidence that shows contamination between mantle and crust. Numbers of xenocrystic zircons are more than members of contemporaneous zircons in rhyodacite, the fact suggests the acid volcanics could be developed from continental curst remelted.

Nd-, Sr-, and Pb- Isotopic Geochemistry

Nd, Sr and Pb isotopic compositions of the bimodal suites from Lower Yangtze region are presented in table 3. The initial Nd and Sr isotopic ratios of basalts and rhyodacites are plotted on $\in Nd(T)$ and initial $^{87}Sr/^{86}Sr$ diagram (Fig. 5). The basalts plot in two clusters, i. e. higher $\in Nd(T)$ group of Calc-alkaline basalts possess $\in Nd(T)=-1\sim-3$. and lower $\in Nd(T)$ group of tholeiitic rocks possess $\in Nd(T)=-6\sim-8$. In contrast, the rhyodacites show very limited $\in Nd$ (T) values ranging $-14\sim-15$. The distributions of basaltic rocks on Fig. 5 are located between subcontinental lithosphere mantle (SCLM) and continental crust $(CC)^{[5,30]}$, that means a possibility mixing SCLM and CC. It is very interested in that the model age T_{DM}^{Nd} of 2288Ma for CB-1 basaltic andesite is almost the same with the U-Pb age of 2218Ma measured from the oldest xenocrystic zircon in the sample. Similarly, T_{DM}^{Nd} age of 1943Ma for XLS-1 rhyodacite is also the same with the U-Pb age of 1949Ma analyzed from the oldest xenocrystic zircon in that sample (See Table 2 and 3). All of the results from Sm-Nd isotopes and U-Pb dating indicate the basaltic magma was prominently contaminated by early Proterozoic crustal rocks, and the early-middle Proterozoic crustal rocks might be the oldest crust fractionated from upper mamtle.

We plot $\in Nd(T)$ vs. $^{206}Pb/^{204}Pb$ diagram for the bimodal suites of Lower Yangtze region (Fig. 6). It shows the location of basaltic samples are between depleted mantle (DM) and continental crust (CC). the calc-alkalic basalts are nearly the extreme of depleted mantle, and tholeiites are transferred to crustal end-member. The distributing array of two basaltic groups on Fig. 6 coincides with those on Fig. 5. Combining above mentioned facts with elemental geochemistry of the basaltic rocks, they indicate the crustal contamination degree in tholeiites is higher than that of calc-alkalic basalts. However, the rhyodacites are products from partial melting of crustal rocks (Fig. 5 and 6).

DISCUSSION

DIfferent Nature of Two Basalt Magma Sources from Hainan Island and Lower Yangtze Region

From the Nd and Sr isotopic data for Hainan Isl. , the basalts appear high $\in Nd$ (T) values from 4. 23 to 7. 56 which could be comparison with those of MORB and IOB. So that, the basalts could come from depleted mantle source. The features of LIL and REE of the basalts support the hypothesis of depleted mantle

Table3. Nd—. Sr— and Pb— isotopic data for the bimodal volcanics from Lower Yangtze region

No.	2	9	12	13	3	5	8	14	15
Sample No.	HS—2	BPQ—5	GD—1	XLS—10	CB—1	CB—4	XXZ	XLS—1	XLS—6
Rock type^d	CAB	CAB	CAB	CAB	BA	TH	BA	RD	RD
Sm (ppm)	3.76	7.23	7.71	6.20	6.85	7.35	6.90	4.08	5.18
Nd (ppm)	19.20	40.20	39.67	32.20	28.58	36.70	30.08	23.01	30.39
$\frac{^{147}Sm}{^{144}Nd}$	0.1185	0.1088	0.1176	0.1165	0.1450	0.1211	0.1391	0.1073	0.1031
$\frac{^{143}Nd}{^{144}Nd}$	0.512452 (11)	0.512343 (19)	0.512201 (19)	0.512138 (10)	0.512117 (9)	0.512154 (8)	0.512178 (18)	0.511686 (9)	0.511738 (8)
$\in Nd(T)^g$	−1.29	−3.13	−6.18	−7.36	−8.62	−7.19	−7.26	−15.91	−14.77
Rb (ppm)	20	34	24.8	24	27.2	24	27.3	82.8	85.8
Sr (ppm)	680	1208	1224	999	292.4	526	414	256	589
$\frac{^{87}Rb}{^{86}Sr}$	0.0851	0.0814	0.0586	0.1307	0.2691	0.1320	0.1908	0.4211	0.9375
$\frac{^{87}Sr}{^{86}Sr}$	0.71275 (4)	0.71793 (4)	0.70887 (3)	0.70744 (2)	0.71455 (5)	0.71040 (5)	0.71175 (2)	0.70970 (2)	0.71366 (4)
$\in Sr(T)$	114.5	188.2	60.7	36.9	131.3	74.7	95.3	30.9	111.6
$\frac{^{206}Pb}{^{204}Pb}$	18.323	17.786	17.833		18.158	17.817	18.117	17.033	
$\frac{^{207}Pb}{^{204}Pb}$	15.463	15.408	15.517		15.520	15.378	15.549	15.388	
$\frac{^{208}Pb}{^{204}Pb}$	37.992	37.836	38.008		38.344	37.860	38.426	37.067	

Note: ⚏CAB—calc—alkalic basalt: BA—basaltic andesite: TH—tholeiite: RD—rhyodacite.
   The values are corrected by ^{146}Nd $^{144}Nd = 0.7219$. 2σ analysis error.
   Using parameters for $\in Nd(T)$: $(^{147}Sm/^{144}Nd)_{CHUR}(0) = 0.1967$; $(^{143}Nd$ $^{144}Nd)_{CHUR}(0)$
   $= 0.512638$; $T = 234Ma$.
   ⚏ Using parameters for $\in Sr(T)$: $(^{87}Rb/^{86}Sr)_{CHUR}(0) = 0.85$; ^{8} Sr $/^{86}Sr)_{CHUR}(0) = 0.7406$;
   $T = 234Ma$.

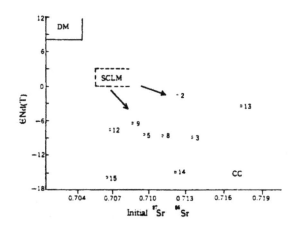

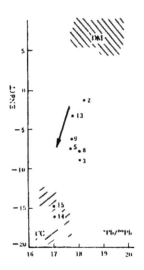

Fig. 5 ∈Nd(T)-Initial ⁸⁷Sr/⁸⁶Sr diagram for the bimodal volcanics from Lower Yangtze region. DM-depleted mantle; SCLM-subcontinental lithosphere mantle; CC-Continental crust. Direction of arrowhead shows the increase of crustal contamination. Sample No. such as those in Table 3.

Fig. 6 ∈Nd(T)-²⁰⁶Pb/²⁰⁴Pb diagram for the bimodal volcanics from Lower Yangtze region. DM-depleted mantle; CC-continental crust; Derection of arrowhead shows the increase of crustal contamination. Sample No. are the same of those in Table 3.

source. Zhu and Wang (1989) studied characteristics of Nd-, Sr-, and Pb- isotopes and geochemistry of Cenozoic basalts from northern Hainan Isl. , where is an one-side of Qiongzhou Strait and belongs to a part of modern rift. The nature of isotopic and elemental geochemistry both ancient basalts of the study and Cenozoic basalts are closely similar to each other[32]. These facts reflect the chemical stability of subcontinental asthenosphere in spatial and temporal two-dimension field.

To compare the mafic volcanics from Lower Yangtze region with those of Hainan Isl. , the earlier Nd, Sr, Pb isotopic data appears lower ∈Nd(T) and lower lead isotopic ratio values. Besides those, we have directly found evidence of old xenocrystic zircons in the basaltic andesite. These could result from large scale contamination of local continental crust. To contrast with the tholeiites, the calc-alkalic basalts from Lower Yangtze region are characterized by higher contents of alkalic elements, lower silica contents, relatively huge values of ∈Nd (T) from −1 to −3 and higher isotopic values of ²⁰⁶Pb/²⁰⁴Pb, ²⁰⁷Pb/²⁰⁴Pb and ²⁰⁸Pb/²⁰⁴Pb ratios. In the geological setting, the tholeiites are located in the centre part of Lower Yangtze rift-depression, and the calc-alkalic basalts are discributed at two outside areas of the location of tholeiites. To combine all above mentioned facts, which show the contamination of continental crust in tholeiitic magma is more serious than that of calc-alkalic basalts.

Chen et al. (1991) measured Nd isotopic compositions of Cenozoic alkalic basalts in Lower Yangtze depression, their ∈Nd(T) values are from 3.3 to 6.4. Among those an alkalic basalt, that was sampled from Fangshan, Jiangning County in our research area, have ∈Nd(T) value of 3. 8. [33], Zhi et al. (1994) reported the same ∈Nd(T) ralues from 4. 2 to 6. 9 and the $^{87}Sr/^{86}Sr$ values from 0. 7034 to 0. 7041 of the Cenozoic alkalic basalts in the same area[34]. The evidence indicates OIB-type mantle exists under the Lower Yangtze Plate. Maybe, we could conclude that the basaltic rocks in the study developed from OIB-type depleted mantle magma source which was contaminated by Pre-Cambrian continental crust[35].

Similar Genesis of Acid Magma Sources for the two bimodal suites

In order to evaluate the potential magma sources for rhyolitic rocks in Hainan Isl. and in Lower Yangtze region, we have to first ascertain the process that best describes the range of isotopic signatures observed. (1) About the rhyolites of bimodal suites from Hainan Isl. , their features of LIL, REE and Nd-, Sr-isotopes are absolutly different from those of basalts of the bimodal suites (Fig. 2 and 3), but these features of rhyolites are very similar to those of Hercynian-Indosinan granitoids, which are huge crust-source type batholithic granites[13,15,25]. And the characteristics of ∈Nd(T) and model ages of T_{DM}^{Nd} of the rhyolites are the same values of Boban gneisses in western Hainan Isl. [24]. All features of the rhyolites observed suggest that they were not derived by fractionation of the basaltic magma. Owing to the eruption of mafic lavas and uplift chambers of the mafic magma, large scale heat flow supported the partial melting of Proterozic continontal crust sources. The rhyolites of the bimodal suites originated from the anatectic melting[10].

(2) In the Lower Yangtze region, the characteristics of elemental and Nd-, Sr-Pb- isotopic geochemistry of rhyodacites are different with the basic volcanics of bimodal suites, and also different from the Cenozoic basalts in the study area. The basic volcanics of bimodal suites are widely distributed over the study area, but their thickness does not exeed 5 metres[28]. So small volume of basaltic rocks that the rhyodacites could not derive by fractionation of the mafic magma. As well as, the $^{143}Nd/^{144}Nd$ values and model age T_{DM}^{Nd} of the rhyodacites are like those Nd isotopic compositions ($^{143}Nd/^{144}Nd = 0. 511774 \sim 0. 512207$; $T_{DM}^{Nd} = 1. 63Ga \sim 2. 12Ga$) of Pre-Permian metamorphic and sedimentary rocks from southern Anhui Prov. where near closely the study area and belongs to an uplift part of Lower Yangtze Plate. [36]. In addition, a lots of xenocrystic zircons have been found in the rhyodacite, their volumes are more than the volumes of contemporaneous zircons. Above mentioned evidence suggests that the magma sources of rhyodacites developed from anatectic melting of continental crust during extension period.

CONCLUSIONS

The results of our geochemical study lead us to the following conclusions regarding the origin of magma in the compositionally two bimodal volcanic fields from continental margin rifts, SE China.

1. The bimodal suites in Hainan Isl. occurred at the transformed stage from intra - to inter- continental rift[10]. It is similar to the thinning crust condition of Red

Sea Rift at present [1,6, ,8]. So the mafic lava wes generated from depleted mantle magma source, that was not contaminated by crust rocks.

2. Another bimodal suites in Lower Yangtze region developed in intra continental rift setting. The big crustal thickness of Lower Yangtze Plate is over 37km. But total thickness of basaltic rocks in the bimodal suites is less than 5 metres. Although OIB type depleted mantle exists as an initial magma source under the plate, the small volume eruption of basaltic rocks associated with weak extension resulted in markedly crustal contamination of mafic magma source during the rifting.

3. Felsic end-members of the two bimodal volcanic rocks, which are from Hainan Isl. and Lower Yangtze region, were generated by anatectic melting of Proterozoic local crust rocks under high heat flow condition in rift zones.

Acknowledgements

The work reported here is partially supported by NSFC grants 4880131 and 49173165. We are grateful to Dr. M. T. McCulloch and RSES ANU proviiding the Sm-Nd isotope lab and financial fund for analyzing samples of Hainan Island in the study. The manuscript was greatly improved following reviews by Prof. D-Z. Wang, Prof. S. Kanisawa and the chief editor of the volume, Prof. Z−N. Li and Dr. Z−C. Zhang are greatly appreciated.

REFERENCES

1. Bohannon, R. G., Naeser, C. W., Schmidt, D. L. and Zimmermann, R. A., The timing of uplift, volcanism and rifting peripheral to Red Sea: A case for passive rifting? J. Geophys. Res., 94, 1683-1701(1989).
2. Leeman, W. P. and Fitton, J. G., Magmatism associated with lithospheric extension: Introduction. J. Geophys. Res., 94, 7682-7684(1989).
3. White, R. and McKenzie, D., Magmatism at rifting zones: The generation of volcanic continental margins and flood basalts. J. Geophys. Res., 94, 7685-7729(1989).
4. Asmeron, Y., Patchett, P. J. and Damon, P. E., Crust-mantle interaction in continental arcs: Inference from the Mesozoic arc in the southwestern United States. Contrib. Mineral. Petrol., 107, 124-134(1991).
5. Sharma, M., Basu, A. R., Cole, R. B. and Decelles, P. G., Basalt rhyolite volcanism by MORB - continental crust interaction: Nd, Sr isotopic and geochemical evidence from southern San Joaquin basin, California. Contrib. Mineral. Petrol., 109, 159 172(1991).
6. Altherr, R., Henjes-Kunst, F. and Baumann, A., Asthenosphere versus lithosphere as possible sources for basaltic magmas erupted during formation of the Red Sea: Constraints from Sr, Pb, and Nd isotopes. E. P. S. L., 96, 259-268(1990).
7. Barrat, J. A., John, B-M., Joron, J. L., Auvray, B. and Hamdi, H., Mantle heterogenity in northeastern Africa: Evidence from Nd isotopic compositions and hydromagmatiphile element geochemistry of basaltic rocks from the Gulf of Tadjoura and southhern Red Sea regions. E. P. S. L., 101, 233-247(1990).
8. Hart, W. K., Wolde-Gabriel, G., Walter, R. C. and Mertzman, A., Basaltic volcanism in Ethiopia: Constraints on continental rifting and mantle interaction. J. Geophys. Res., 94, 7731-7748(1989).
9. Wang, H-Z. and Mo, X-X., An outline of the tectonic evolution of China. Episodes, 18(1/2), 6-16(1995).

10. Xia. B-D. , Shi. G-Y. , Fang. Z. , Yu. J-H. , Wang. C-Y. , Tao. X-C. and Li.H-M. , The Late Palaeozoic rifting on Hainan Island. China. Acta Geologica Sinica. 4(4). 341-355 (1991).

11. Gu. L-X. and Xu. K-Q. , On the Carboniferous submarine massive sulphide deposits in the lower reaches of the Changjiang (Yangtze) River. Acta Geologica Sinica. 60. 176-188 (1986)(in Chinese with English abstract).

12. Xia. B-D. , Liu. H-L. , Fang. Z. and Lu. H-B. , Sedimentary deposits of Longtan Group and Late Palaeozoic tectonic framework in the eastern Yangtze Plate. China. Acta Sedimentologica Sinica. 10.23-34(1992)(in Chinese with English abstract).

13. Fang. Z. , Geochronology and Sr. O isotope studies of granites in Hainan Island. China. Chem. Geol. . 70.20(1988).

14. Wang. X-F. , Ma. D-Q. and Jiang. D-H. , Geology of Hainan Island (I); Igneous Rocks. Geological Publishing House. Beijing. 6-46(1991)(in Chinese).

15. Fang. Z. , Xia. B-D. , Yu. J-H. , Tao. X-C. , Yang. J-D. and Li. H-M. , Rb-Sr geochronological and oxygen isotopic studies of the Palaeozoic rocks in Hainan Island. Acta Petrologica Sinica. 3. 296-302(1992) (in Chinese with English abstract).

16. Scientific Research Team of High-grade Iron Deposit. Chinese Academig of Science. Geology of Hainan Island and Geochemistry of the Shilu Iron Deposit. Science Press. Beijing. 376pp(1986) (in Chinese).

17. Fang. Z. , Tao. X-C. , Xia. B-D. and Yang. J-D. , Isotopic record on Palaeozoic crust evolution within Hercynian rift in Hainan Island. China. Geol. Soc. Aust. Abstr. . 27. 33 (1990).

18. Xia. B-D. , Yu. J-H. , Fang. Z. , Wang. C-Y. and Shi. G-Y. , Carboniferous bimodal volcanic rocks and their plate tectonic setting. Hainan Island. Chinese J. of Geochemistry. 11 (1). 70-79(1992).

19. Meschede. M. , A method of discriminating between different types of mid-ocean ridge basalts and continental tholeiites with the Nb-Zr-Y diagram. Chem. Geol. . 567. 207-218 (1986).

20. Pearce. J. A. and Norry. M. J. , Petrogenetic implications of Ti. Zr. Y and Nb variations in volcanic rocks. Contrib. Mineral. Petrol. . 69. 33-47(1979).

21. Sounders. A. D. , The rare-earth element characteristics of igneous rocks from the ocean basins. In P. Henderson (ed.). Rare-Earth Element Geochemistry. Elsevier Sci. Pub. Co. . 205-230(1984).

22. Xia. B-D. , Yu. J-H. , Fang. Z. , Wang. C Y. and Chu. X-J. , Geochemical characteristics and origin of the Hercynian-Indosinian granites of Hainan Island. China. Geochimica. 4.365-373(1990) (in Chinese with English abstract).

23. Henger. E. and Pallister. J. S. , Pb. Sr and Nd isotopic characteristics of Tertiary Red Sea Rift volcanics from the Central Saudi Arabian coastal plain. J. of Geophys. Res. . 94. 7749-7755(1989).

24. Chen. H-F. , The discovery of Precambrian crystal basement and komatiite in Hainan Island. Tectonics and Minerogenesis. 15(1). 62(1991)(in Chinese).

25. Fang. Z. , Yu. J-H. , Xia. B-D. , Tao. X-C. , Li. H-M. and yang. J-D. , Sm-Nd isotopic characteristics of Hercynian-Indosinian granites and xenoliths in Hainan Island. S. China. J. of Nanjing University (Natural Sci. Edition). 31(2). 338-343(1995) (in Chinese with English abstract).

26. Xia. B-D. , Zhong. L-R. , Fang. Z. and Lu. H-B. , Early Permian Gufengian argillized volcanics in the Lower Yangtze region. Geological Review. 40(1). 64-73(1994)(in Chinese with Enlgish abstract).

27. Xia. B-D. , Zhong. L-R. , Fang. Z. and Lu. H-B. , The origin of bedded cherts of the Ear ly Permian Gufeng Formation in the Lower Yangtze area. E. China. Acta Geologica Sinica. 8 (4). 372-386(1995).

28. Fang. Z. , Xia. B-D. , Chu. X-J. and Liu. S-H. , Nwely discovered bimodal subvolcamic rocks in the late Permian from S. Jiangsu Province and single zircon U-Pb geochronology. J. of Stratigraphy. 18(3). 168-172(1994)(in Chinese with English abstract).

29. Wilson. M. , Igneous Petrogenesis; A global tectonic approach. Academic Davison of Unwin Hyman Ltd. . London. 460(1989).

30. Weaver. B. L. and Tarney. J.. Chemistry of the subcontinental mantle: Inference from Archaean and Proterozoic dykes and continental flood basalts. in: C. J. Hawkesworth and M. J. Norry (ed.) Continental Basalts and Mantle Xenoliths. Shiva Publishing Ltd. . Cheshire. 209-229(1983).

31. Taylor. S. R. and McLennan. S. M.. The Continental Crust: Its compositon and evolution. Blackwell. Oxford. 312 pages(1985).

32. Zhu. B-Q. and Wang. H-F.. Nd-Sr-Pb isotopic and chemical evidence for the volcanism with MORB-IOB source characteristics in the Lei-Qiong area. China. Geochemica. 3. 193-201(1989) in Chinese with English abstract).

33. Chen. D-G. . Yang. J-D. and Wang. Y-X.. Nd isotopic compositon and significance of some Cenozoic volcanic rocks from Jiangsu. Anhui and Shandong. China. Chinese Sci. Bull. . 36(10). 843-846(1991).

34. Zhi. X-C. . Chen. D-G. . Zhang. Z-Q. and Wang. J-H.. Nd and Sr isotopic compositions for Later Tertiary alkalic basalts from Liuhe-Yizheng. Jiangsu Prov. . China. Acta Petrologica Sinica. 10(4). 382-389(1994)(in Chinese with English abstract).

35. Heaman. L. M. and Machado. N.. Timing and origin of midcontinent rift alkaline magmatism. North America: Evidence from the Coldwell Complex. Contrib. Mineral. Petrol. . 110. 29-303(1992).

36. Chen. J-F. . Zhou. T-X. . Xing. F-M. . Xu. X. and Foland. K. A.. Provenances of low grade metamorphic and sedimentary rocks from southern Anhui Prov: Evidence of Nd isotope compositions. Chinese Sci. Bull. . 35(9). 747-750(1990).

Proc. 30th Int'l. Geol. Congr., Part 15 pp. 39–49
Li et al. (Eds)

PGE and Au Abundances in MORB Glasses from Depleted Mantle Melting and in Alkali Basalts from Metasomatized Mantle Melting.

GERALD HARTMANN
Geochemisches Institut der Universität, Goldschmidtstr. 1, D-37077 Göttingen, Germany

Abstract

MORB glasses from the East Pacific Rise and the Lau Basin are distinguished from alkali basalts from Germany by low absolute concentrations of all PGEs and an overall unfractionated to moderately fractionated PGE distribution pattern, with a ratio (PPGE/ IPGE)mn of about 1 (to 10). The whole MORB data set indicates a relatively narrow range of PGE contents (Ir 0.01-0.09 ppb, Ru 0.02-0.1 ppb, Rh 0.01-0.03 ppb, Pt 0.05-0.4 ppb, Pd 0.03-0.4 ppb) with Au varying from 0.08 to 10 ppb. In contrast, all alkali basalts display a fractionated PGE-pattern, (PPGE/ IPGE)mn > 1. The absolute contents of all PGEs and Au increase from continental quartz tholeiites, which are similar to the low MORB values, alkali olivine basalts to olivine nephelinites. The non-metasomatized and on the other side the metasomatized and probably more oxidized source mantle region have two fundamentally contrasting patterns of PGE behaviour and fractionation of the basalts. It is assumed that the MORB glasses were generated by the separation of a sulphur-saturated melt from a non-metasomatized MORB-source peridotite. A residual sulphide phase was retained in the mantle source during partial melting and these sulphides have mainly retained the precious metals. Because sulphide-melt/ silicate-melt partition coefficients for the different PGEs are not significantly different from one another, unfractionated PGE-pattern and low concentrations were consequently determined in the MORB glasses. In some cases, subsequent post-melting segregation and assimilation of chromites and olivines can lead to a progressively PGE fractionation in MORB. In marked contrast to ocean floor tholeiites, oxidizing metasomatic fluids that have conditioned the lithospheric upper mantle for the formation of alkali basaltic magmas have also oxidized and dissolved the PGE-concentrating sulphides in the source mantle. The oxidation of the mantle sulphides by metasomatic processes gave rise to a redistribution between the residual mantle minerals of the PGEs and a change in their geochemical behaviour. On account of their higher solubility in silicate melts PPGEs and Au are favourably transported along with sulphur into the melt, while the IPGEs are retained in the residue probably incorporated into spinels or precipitated as alloys, possibly engulfed by recrystallizing spinels. Therefore mantle metasomatism is one of the most essential processes which indirectly dominate the PGE distribution between magma and peridotitic residue. The sulphide oxidizing and dissolving character of the mantle metasomatism is the triggering factor, ultimately responsible for inter-element fractionation and (PPGE/ IPGE)mn > 1 in products of e. g. the continental intra-plate volcanism of the Hessian Depression. The influence of the degree of partial melting on the PGE distribution is of minor importance, which is demonstrated by the fact that the absolute concentrations of all precious metals in the investigated set of basalts decrease with increasing degree of melting.

Keywords: platinum-group elements, gold, MORB, alkali basalt

INTRODUCTION

In order to highlight the principal behaviour of the highly siderophile platinum-group elements (IPGE: Ir, Ru; PPGE: Rh, Pt, Pd) and Au during partial melting and mantle metasomatism a wide variety of asthenospheric and lithospheric ultramafic and mafic rocks were investigated [1, 2]. The chemical composition of the investigated mafic rocks ranges from MORB glasses to different alkali basalts. The combined investigation of asthenospheric MORB-producing ultramafic rocks and mid-ocean ridge basalts offers the opportunity to study the behaviour of the highly siderophile elements and their fractionation during partial melting processes which took place in a mildly depleted, unmetasomatized upper mantle by moderate degrees of melting in a sulphide-saturated source region [3].

In addition to this process of depleted mantle melting, mafic products and ultramafic residues of metasomatized mantle melting were investigated. According to Wedepohl et al. [4, 5] the continental intra-plate volcanism of the Hessian Depression in Germany yields varies types of alkali basalts whose partial melting events were preceded by mantle metasomatism.

The two groups of mafic rocks are regarded as approaching the criteria for primary magmas, notably by their high Mg#s, and Cr and Ni contents.

ANALYTICAL TECHNIQUE

The precious metals were determined by ICP-mass spectrometry after preconcentration with nickel sulphide fire-assay and tellurium coprecipitation. To overcome the difficulties which are caused by the so called nugget-effect and to improve the overall accuracy of the analysis, the NiS fire-assay method was arranged to sample sizes of up to 500 g for basalts. Reagents (120 g Na_2CO_3, 440 g $Na_2B_4O_7$, 9 g S, 120 g SiO_2 and 15g refined Ni) and the sample (400-500 g) were directly weighed into a 3 l-glass bottle and thoroughly mixed. After transferring the flux in a 3 l-fire assay crucible, the mixture is fused at 1050°C for 24 hours. The assay bead is dissolved in conc. HCl. The solution is diluted by distilled H_2O and dissolved fractions of noble metals are coprecipitated by coagulating tellurium. The precipitate is collected onto a cellulose nitrate membrane-filter which is subsequently dissolved by aqua regia in a FEP-flask and the solution is diluted with distilled water to give a final acid concentration of approx. 5 %. The instrument employed for measurement was an Elan 5000 model from Perkin Elmer. The whole technique is comparable to that of Robert et al. [6], Fryer and Kerrich [7] and Jackson et al. [8]. The standard deviation of the method at the PGE concentration level of basalts varies between 20 % and 50%. The accuracy of the method was controlled by analysing international reference rocks from the Geological Survey of Canada/ Mineral Resources Divisions and the Council for Mineral Technology/ South Africa.

HIGHLY SIDEROPHILE ELEMENT ABUNDANCES IN MORB GLASSES

The investigated MORB glasses from the East Pacific Rise and the northern Lau Basin show N-MORB type characteristics in major and trace elements. Isotopic compositions vary within a range typical for N-MORB [9, 10].
The MORB glasses from the East Pacific Rise and the Lau Basin are distinguished by low absolute concentrations of all PGEs. Additionally, the whole MORB data set indicates a relatively narrow range of PGE contents (Ir 0.01-0.09 ppb, Ru 0.02-0.1 ppb, Rh 0.01-0.03 ppb, Pt 0.05-0.4 ppb, Pd 0.03-0.4 ppb). The individual values don't deviate from the averages by more than a factor of 5. This tight clustering of the PGE's around the averages in MORB is one of the most significant aspect of the present investigation. This is confirmed by the few available PGE-MORB data from the literature for the Indian Ocean and the Mid-Atlantic Ridge. The PGE data agree very well with the few previously published MORB values [11, 12], apart from the anomalously high PPGE content which Hertogen et al. [12] determined in two atypical MORB samples. In contrast, the Au contents vary around the low side but also exceed the high side of the previously existing MORB data range, with values varying from 0.1 to 10 ppb. Because in fertile to moderately mantle sections Au is homogenously distributed with an average value of 1 ppb, these variable concentrations in MORB glasses probably reflect a greater mobility of Au during melting processes, as was previously deduced from our investigations of ultramafic rocks.

Table 1. Average platinum-group element and gold abundances in MORB glasses and mafic rocks from the intra-plate volcanism of the Hessian Depression in Germany

ppb	MORB East Pacific Rise n= 4	MORB Northern Lau Basin n= 5	Continental quartz tholeiites n= 6	Alkali olivine basalts n= 6	Olivine-nephelinites, nepheline-basanites n= 4	Evolved olivine-nephelinites, nepheline basanites
Ir	0.04	0.028	0.03	0.13	0.62	0.065
Ru	0.05	0.033	0.045	0.22	1.0	0.1
Rh	0.01	0.012	0.01	0.06	0.23	0.025
Pt	0.07	0.142	0.07	0.48	1.79	0.210
Pd	0.06	0.113	0.063	0.42	1.74	0.180
Au	2.5	2.13	0.675	0.41	0.7	1.26

MORB/ (Northern Lau Basin)

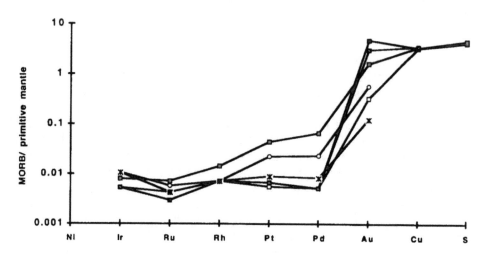

MORB/ East Pacific Rise

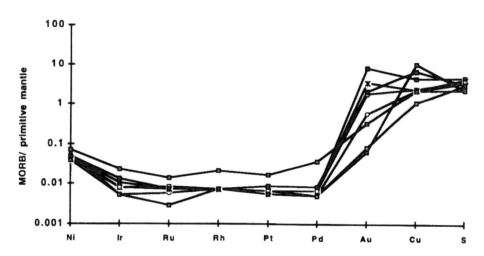

Figure 1. PGE and Au distribution patterns of MORB glasses normalized with the concentrations of the primitive mantle according to Hartmann (1996c).

The overall assumption, considering all investigated MORB samples from different geographic locations, is that ocean floor basalts are derived from an upper mantle section that is fairly homogeneous in respect to its precious metal abundances. This assumption confirms our previous results obtained from investigations of worldwide upper mantle rocks.

The investigated MORB glasses are marked by an overall unfractionated to moderately fractionated PGE distribution pattern, with a ratio of (PPGE/IPGE)mn of about 1.

Because the highly siderophile elements are concentrated in the upper mantle sulphides their distribution between MORB source peridotites and related MORB melts depends mainly on the sulphur solubility of the separated melt and the dominating sulphur species involved in the partial melting process. The determined relatively reduced redox state of asthenospheric MORB-producing peridotites from the Ivrea-Zone with an oxygen fugacity of Δlog $f O_2$ around -1 below the FMQ-buffer [1, 2], is evidence for a melting process in which mantle sulphides were sustained during partial melting and therefore dominate the melt-vapor equilibrium. The model for the distribution of nobel metals between MORB-source peridotites and MORB melts is based on high and similar sulphide melt-silicate melt partition coefficients for these elements [13, 14]. The mantle sulphides are in liquid state at the incipient stage of partial melting events on account of their comparatively low melting range, (see for instance Lorand et al., 15). In the MORB-producing, relatively reduced environment the mantle sulphides dissolve in the silicate melt, and release the sulphide-forming elements S, Cu, Fe and Ni and the accomodated PGEs and Au. As melting proceeds the sulphides dissolve at the level of sulphur saturation in proportion to the degree of melting and therefore release the PGEs and Au in this proportion. This is suggested by correlations between Al_2O_3, sulphur and the PGE contents of the residual, MORB-producing mantle which varies in relation to the amount of partial melt extracted [3]. The incompatible behaviour of all precious metals depends on the involvement of mantle sulphides in partial melting processes. As the degree of melt separation increases the content of sulphur and all precious metals decreases in the source [3]. But a residual sulphide phase is retained in the mantle source during partial melting. These sulphides retain a related part of the precious metals with an unfractionated PGE-pattern [3], because sulphide melt /silicate melt partition coefficients for the different PGEs are not significantly different from one another. According to Hartmann [3], MORB-source peridotites are characterized by moderately depleted, unmetasomatized, sulphur-containing lherzolites with an unfractionated PGE-distribution pattern, (PPGE/ IPGE)mn = 1, and absolute PGE concentrations of about 0.8 % of the chondritic concentrations. Primary MORB melts should exhibit the same PGE distribution pattern as the sulphide melt at the time of sulphur saturation of the silicate melt. If secondary alteration, as chromite segregation or contamination do not occur during ascent the magma should preserve an unfractionated PGE-pattern. Unfractionated distribution patterns, with ratios of (PPGE/ IPGE)mn of about 1, were observed in a selection of samples from the East Pacific Rise MORB besides fractionated PGE distribution patterns. In some cases, subsequent post-melting segregation and assimilation of sulphides, chromites and olivines can lead to a progressively PGE fractionation in MORB.

HIGHLY SIDEROPHILE ELEMENT ABUNDANCES IN ALKALI BASALTS

The continental intra-plate volcanism of the Hessian Depression in Germany is represented by various types of alkali basalts whose partial melting events were preceded by mantle metasomatic processes, according to Wedepohl et al. [5]. The Mg number of selected alkali olivine basalts, nepheline basanites and olivine nephelinites approaches that of primary basaltic melts, while the quartz tholeiites are derivative magmas. The chemical compositions of the primary melts indicate a generation by decompressional partial melting of metasomatically altered asthenospheric mantle [4].

The absolute PGE concentrations, and the PGE distribution patterns of the continental tholeiites are basically the same as those of ocean floor tholeiites. In contrast, the content of all PGEs and Au increases from continental quartz tholeiites with a total amount of 0.2 ppb PGEs through alkali olivine basalts with a total amount of 1.3 ppb PGEs to olivine nephelinites and nephelin-basanites with a total amount of 5.4 ppb PGEs. The average Pd concentration increases in this rock sequence from 0.06 ppb through 0.4 ppb to 1.8 ppb. In marked contrast to continental and ocean floor tholeiites, all alkali basalts display a fractionated PGE-pattern, with a ratio (PPGE/ IPGE)mn > 1. These basalts show Pd/Ir ratios of about 3 for ON/ NB and 4 for AOB. It is possible to group the various mafic rocks not only according to their total PGE content but also to their concentrations of lithophile incompatible elements, for example LREE, which increase in the same rock sequence from tholeiites through AOB to ON and NB.

In order to investigate the influence of metasomatic processes on the distribution of the PGEs between ultramafic and mafic rocks, in addition to the basalts ultramafic tectonites and xenoliths have been investigated which were secondarily altered by fluids in open-system processes. According to Hartmann [1, 2] mantle metasomatism does not directly influence the PGE abundance of the upper mantle but can affect the Au, Cu and S abundances. Mantle metasomatism does not cause an enrichment of the PGEs in residual peridotites. ^{57}Fe-Mössbauer measurements indicated higher Fe^{3+}- concentrations in spinels of metasomatically overprinted samples than in fertile asthenospheric rocks. These peridotites display an oxidation state with $\Delta \log f O_2 = +1.2$ above the FMQ buffer [1, 2]. Oxidizing metasomatic fluids that have conditioned the lithospheric upper mantle for the formation of alkali basaltic magmas [16, 17] have also oxidized and dissolved the PGE-concentrating sulphides in the source mantle. The metasomatic fluid-rock

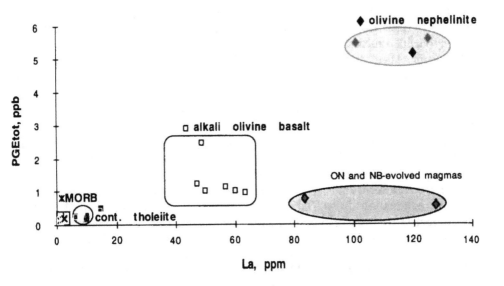

Figure 2. The average total PGE content increases with the La content from MORB through continental tholeiites to alkali basalts.

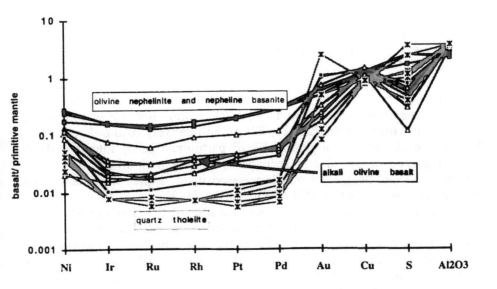

Figure 3. PGE and Au distribution pattern in alkali basalts from metasomatized mantle melting.

interaction by oxidizing fluids causes the oxidation of mantle sulphides and thus the sulphur was probably mainly dissolved as sulphate in the silicate melt during partial melting. Evidence for such an oxidation is the occurrence of sulphate in alkali basaltic rocks from this type of mantle [18]. The transition in the redox state from a relatively reduced MORB-producing mantle to an relatively oxidized mantle causes a transition in the sulphur modification from S^{2-} to SO_4^{2-}. This alteration has a distinct impact on the distribution of the precious metals between residuum and melt, because the PGEs lost there original host mineral. The oxidation of the mantle sulphides by metasomatic processes forced the PGEs to a redistribution between the residual mantle minerals and a change in their geochemical behaviour. As indicated by the distribution patterns of the metasomatized ultramafic rocks, at the time of sulphide exhaustion up to 70 % of the original IPGE and 20 % of the PPGE (in comparison to the primitive mantle) must have been accomodated in non-sulphide phases [1, 2]. The PGE distribution patterns of such almost sulphide-free, subcontinental lithospheric mantle rocks from the Hessian Depression (Germany), Eifel (Germany), Massif Central (France), Kilbourne Hole (USA), San Carlos (USA), China and Antartica are marked by a strong depletion of the PPGEs, but only a minor depletion of the IPGEs, in comparison to the sulphide-containing peridotites [1, 2]. The metasomatized peridotites with geochemical features of secondary alteration by modal and kryptic mantle metasomatism have PGE distribution patterns with a ratio (PPGE/ IPGE)mn < 1 [1, 2]. The IPGEs were retained in the residue due to their minor solubility in silicate melts in comparison to the PPGEs. One residual mantle mineral in sulphur-depleted peridotites which can accomodate appreciable amounts of IPGEs may be spinel. Spinel/ silicate partition coefficients indicate that the IPGE are moderately compatible while the PPGE are incompatible for spinel, according to Capobianco et al. [19]. It could be posssible that simultaneously with the oxidation of the sulphides, high-melting Os-Ir alloys were formed which would support an additional fixation of the IPGE in the residue. The abundance of the PPGEs in these residual peridotites were seriously affected by mantle metasomatism followed by partial melting. Because of their higher solubility in silicate melts [20, 21, 22] at the relatively oxidized redox state (and in the absence of sulphides) PPGE partitioned partly into the silicate melt. Therefore depleted and secondarily altered peridotites, affected by various types and degrees of mantle metasomatism, display fractionated PGE distribution patterns, with (PPGE/ IPGE)mn < 1 [1, 2].

The sulphide oxidizing and dissolving character of this mantle metasomatic process is ultimately responsible for the inter-element fractionation and a (PPGE/ IPGE)mn > 1 in products of e. g. the continental intra-plate volcanism of the Hessian Depression. The overall positive correlations between the [87]Sr/[86]Sr-, (La/Yb)mn-, (Pd/Ir)mn-ratios and the absolute PGE concentrations in the investigated basalts imply that the PGE abundance and fractionation of the melts increase with increasing degree of mantle metasomatism in partial melting. This means in fact, that with increasing $f O_2$

of the melting environment the solubility of the PGEs, especially of the PPGEs, in the melt increases. The influence of the degree of partial melting on the PGE distribution is of minor importance, which is demonstrated by the fact that the absolute concentrations of all precious metals in the investigated set of basalts decrease with increasing degree of melting, from MORB magmas to ON.

Nevertheless, the content of all PGEs in evolved nephelinites tends to be lower than in the primary rocks (see Table 1) accompanied by varying (Cu/ Pd)mn and (Ni/ Ir)mn ratios, suggesting that in evolved magmas the noble metals were depleted by sulphides. This is evidence for secondary processes which affect the PGE distribution in basalts.

Acknowledgements

I thank Prof. Dr. K. H. Wedepohl for his helpful review of an early version of the manuscript. Dr. W. Bach of the GeoForschungs Zentrum Postdam, Dr. von Stackelberg and Dr. V. Marchig of the Federal Geological Survey (BGR) at Hannover kindly provided MORB glasses for this investigation. U. Siewers and H. Lorenz of the Federal Geological Survey (BGR) at Hannover offered assistance and the facilities for ICP-MS measurements. I would like to thank Mrs. A. Reitz for her technical assistance. This investigation was supported by grants from the Deutsche Forschungsgemeinschaft (DFG).

REFERENCES

1. G. Hartmann. Häufigkeit und Verteilung der Platin-Gruppenelemente und des Goldes im Erdmantel und der Erdkruste. Thesis of Habilitation, University of Göttingen (1996a).
2. G. Hartmann. Platinum-group element and gold abundances in the asthenospheric and lithospheric mantle and their behaviour during partial melting and mantle metasomatism (submitted to Naturwissenschaften) (1996b).
3. G. Hartmann. The abundance and distribution of highly siderophile elements in fertile sulphide-bearing peridotites. Constraints on the primitive mantle (submitted to Jour. of Petrology) (1996c).
4. K. H. Wedepohl, E. Gohn and G. Hartmann. Cenozoic alkali basaltic magmas of western Germany and their products of differentiation. *Contrib. Mineral. Petrol.* **115**, 253-278 (1994).
5. K. H. Wedepohl. Origin of the Tertiary basaltic volcanism in the northern Hessian Depression. *Contrib. Mineral. Petrol.*, **89**, 122-143 (1985).
6. R. V. D. Robert, E. van Wyk and R. Palmer. Concentration of noble metals by a fire-assay technique using nickel sulphide as the collector. Nat. Met. Rep., 1371, Johannesburg (1971).
7. B. J. Fryer and R. Kerrich. Determination of precious metals at ppb levels in rocks by a combined wet chemical and flamless atomic absorption method. *Atomic Absorption Newsletter* **17**, 4-6 (1978).
8. S. E. Jackson, B. J. Fryer, W. Gosse, D. C. Healey, H. P. Longerich and D. F. Strong. Determination of the precious metals in geological materials by inductively coupled plasma-mass spectrometry (ICP-MS) with nickel sulphide fire-assay collection and tellurium coprecipitation. *Chem. Geology* **83**, 119-132 (1990).
9. W. Bach, E. Hegner, J. Erzinger and M. Satir. Chemical and isotopic variations along the superfast spreading East Pacific Rise from 6 to 30° S. *Contrib. Mineral. Petrol.* **116**, 365-380 (1994).
10. G. Loock, W. F. McDonough, S. L. Goldstein and A. W. Hofmann. Isotopic composition of volcanic glasses from the Lau Basin. *Marine Mining* **9**, 235-245 (1990).
11. C.-L. Chou, D. M. Shaw and J. H. Crocket. Siderophile trace elements in the earth's oceanic crust and upper mantle. *J. Geophys. Res* **88**: A507- A518 (1983).
12. J. Hertogen, M.-J. Janssens and H. Palme. Trace elements in ocean ridge basalt glasses: implications for fractionations during mantle evolution and petrogenesis. *Geochim. Cosmo. Acta* **44**, 2125-2143 (1980).
13. C. L. Peach, E. A. Mathez and R. R. Keays. Sulphide melt-silicate melt distribution coefficients for noble metals and other chalcophile elements as deduced from MORB: Implications for partial melting. *Geochim. Cosmo. Acta* **54**, 3379-3389 (1990).

14. C. L. Peach, E. A. Mathez, R. R. Keays and S. J. Reeves. Experimentally determined sulphide melt-silicate melt partition coefficients for iridium and palladium. *Chem. Geol.* **117**, 361-377 (1994).

15. J. P. Lorand, R. R. Keays and J. L. Bodinier. Copper and noble metal enrichments across the lithosphere-asthenosphere boundary of mantle diapirs: Evindence from the Lanzo lherzolite massif. *Jour. Petrol.* **34**, 1111-1140 (1993).

16. G. Hartmann and K. H. Wedepohl. Metasomatically altered peridotite xenoliths from the Hessian Depression (Northern Germany) *Geochim. Cosmochim. Acta* **54**, 71-86 (1990).

17. G. Hartmann and K. H. Wedepohl. The composition of peridotite tectonites from the Ivrea complex (N-Italy): Residues from melt extraction. *Geochim. Cosmochim. Acta.* **57**, 1761-1782 (1993).

18. Y. Muramatsu and K. H. Wedepohl. Chlorine in Tertiary basalts from the Hessian Depression in NW Germany. *Contrib. Mineral. Petrol.* **70**, 357-366 (1979).

19. C. J. Capobianco, R. L. Hervig and M. J. Drake. Experiments on crystal/ liquid partitioning of Ru, Rh and Pd for magnetite and hematite solid solutions crystallized from silicate melt. *Chem. Geol.* **113**, 23-43 (1994).

20. A. Borisov and H. Palme. The solubility of Ir in silicate melts: new data from experiments with Ir10Pt90 alloys. *Geochim. Cosmochim. Acta* **59**, 481-487 (1995)

21. A. Borisov, H. Palme and B. Spettel. Solubility of palladium in silicate melts: Implications for core formation in the Earth. *Geochim. Cosmochim. Acta* **58**, 705-716 (1994).

22. H. St. C. O'Neill, D. B. Dingwell, A. Borisov, B. Spettel and H. Palme. Experimental petrochemistry of some highly siderophile elements at high temperatures, and some implications for core formation and the mantle's early history. *Chem. Geology* **120**, 255-273 (1995).

Proc 30ᵗʰ Int'l. Geol. Congr., Part 15, pp. 51 – 64
Li et al.(Eds)

The Role of Sulphide-Carbonate-Silicate and Carbonate-Silicate Liquid Immiscibility in the Genesis of Ca-Carbonatites

L.N.KOGARKO
Vernadsky Institute of Geochemistry, Moscow

Abstract

During the investigation of a harzburgite nodule from the Montana Clara Volcano (Canary island archipelago) the evidence of the primary carbonate melt was discovered. This carbonate is enriched in calcium and it occurs together with glass containing sulphide globules. In addition to primary olivine and orthopyroxene there are pockets of fine-grained minerals belonging to the metasomatic second generation (more magnesium olivine, sodium-bearing clinopyroxene, less aluminous spinels). The metasomatic assemblage was formed by reaction of sodium-bearing dolomitic melt with the harzburgite according to the reactions:

$2MgSiO_3 + CaMg(CO_3)_2 \quad 2Mg_2SiO_4 + CaMgSi_2O_6 \quad 2CO_2$

$3CaMg(CO_3)_2 + CaMgSi_2O_6 = 4CaCO_3 + 2Mg_2SiO_4 + CO_2$

The calciocarbonatite and sulphide phase almost invariably form globules in the silicate glass indicating the existence of three immiscible liquids under upper mantle conditions resulting from melting of the metasomatised mantle material during the uprising and adiabatic decompression. Our experiments reveal carbonate-silicate-sulphide immiscibility. Therefore the investigated mineral assemblage including carbonate and glass can be considered as a micro model of the generation of the Ca-rich carbonatitic magmas during the processes of the partial melting of carbonatized merasomatized oceanic mantle. The development of carbonatite magmatism on Canary Islands (Lanzarote), Cape Verde Islands is likely to be related to the partial melting of carbonatized mantle in the South Atlantic, which took place over a vast territory.

Another example of carbonate-silicate immiscibility under crustal conditions is represented by phonolites (Polar Siberia, Maimecha-Kotui) containing carbonate globules. We investigated some dykes of carbonatitic massif Dolbykha which comprise olivine and melilite nephelinites, nosean, calcite and cancrinite phonolites, calcite trachytes and calcite carbonatites. Some ultra alkaline phonolitic dykes contain carbonate-bearing globules.Globules consist of polycrystalline calcite aggregate and contain albite, mica, apatite, Sr-lueshite, zircon, ancylite, ilmenite and strontianiteT. There are phenocrysts of albite, mica and ilmenite in phonolites. There are also albite, mica, calcite and nepheline present in the groundmass. The analysis of these materials in the light of experimental data on the liquid immiscibility in carbonate-silicate systems suggests that the separation of carbonatite melts from phonolitic ones took place due to the immiscibility in liquid state. We propose that originally carbonate melts containe significantly higher alkali concentrations which were subsequently lost in fluid phase due to incongruent dissolution of calcium-sodium carbonates in aqueous fluid at low temperatures. Our discovery of nyerereite in carbonatite of Polar Siberia confirms this assumption.

Thus one of the very important mechanisms of the genesis of Ca-rich carbonatite melts was the formation of liquid immiscibility which may take place in mantle or crustal conditions.

Keywords: liquid immiscibility, carbonatites, mantle, phonolites, Montana Clara, Polar Siberia

INTRODUCTION

At present three groups of carbonatite origin models are generally discussed:1 - carbonatites may be formed during the partial melting of carbonated mantle: 2 -

Recently the detailed studies of the geochemistry of mantle nodules and primary mantle carbonates permitted to conclude that carbonatite magmas may be generated in the mantle (Amundsen, 1987; Pyle and Haggerty, 1994; Kogarko et al., 1995a). Experimental data (Wyllie, 1989; Wyllie et al. 1990; Eggler,1989) support the possibility of the generation of carbonatites in the processes of partial melting of carbonated mantle substrate. Mineralogical and geochemical investigations of some carbonatitic complexes (Bailey,1989; Kogarko, 1993) are also in agreement with these ideas.

The experimental investigations conveyed by many authors suggest that the liquid immiscibility plays the leading role in the genesis of carbonatites (Freestone and Hamilton,1980; Hamilton et al.,1989; Kjarsgaard and Peterson, 1991). Experimental data demonstrated the presence of the extensive fields of two immiscible liquids in system $SiO_2+Al_2O_3+TiO_2-MgO+FeO-CaO-Na_2O+K_2O-CO_2$. The immiscibility gap expands with increasing pressure and alkalis content and according to Woh-jer and Wyllie (1996) with the decrease in magnesium concentration. From these data a conclusion was drown that residual magmas during differentiation enriched in alkalis may intersect the boundary of silicate-carbonate liquid immiscibility gap under crustal conditions. However, recently Woh-jer and Wyllie (1994, 1995) discussed a number of limitations on the hypothesis of the genesis of carbonatites by liquid immiscibility and they maintained that beyond any doubts there is no single process responsible for the formation of all carbonatites. According to their opinion fractional crystallization of carbonate-bearing alkaline magmas might be very important process of carbonatites origin. Similar ideas were proposed by Twyman and Gittins (1987).

Because the problem of the genesis of carbonatites still remains unsolved the field evidence, geological and mineralogical observations on the interrelationships between carbonate and silicate material in magmatic rocks become very important.

In this paper we investigate the metasomatic interaction between sodium- and Ca-rich carbonate liquid with mantle material from Montana Clara Island (Canary archipelago) resulted in the wehrlitization and carbonatization of primary harzburgite. The partial melting of this material leads to the formation of silicate-carbonate-sulphide liquid immiscibility and generation of Ca-rich carbonatitic melts. We have investigated the immiscibility in the system Ca-rich carbonate-Fe. Ni sulphide-silicate melt of phonolitic composition containing F. The immiscibility has been observed in the investigated system exhibited in the complete separation of carbonate and silicate liquids whereas sulfide melt was present in the form of globules in both liquids. Therefore the investigated mantle material can be considered as a micro model of the generation of the Ca-rich carbonatites during the processes of the partial melting of wehrlitized and carbonatized mantle.

Moreover we studied carbonate-silicate liquid immiscibility phenomena in phonolitic dyke from Maimecha - Kotui province (Polar Siberia) containing carbonate globules. The interrelationships between silicate and carbonate materials in this phonolite and compositions of minerals testify to the immiscibility phenomena in these rocks. Recently Peterson (1989), Kjarsgaard and Peterson (1991) reported the possibility of the generation of alkali-poor carbonatites of Shombole Volcano (East Africa) as a result of silicate-carbonate immiscibility. This study was based on the petrographic and experimental observations of nephelinites, containing carbonate globules. In this paper we summarize data on petrography, mineral compositions of carbonate-bearing phonolite represent quench droplets of carbonate immiscible melt which separated in intermediate magmatic chamber during the differentiation of alkaline magma. The liquid immiscibility is strongly controlled by CO_2 pressure (depth of magma chamber).

PARTIAL MELTING OF CARBONATED PERIDOTITE AND CARBONATE-SILICATE-SULFIDE LIQUID IMMISCIBILITY IN THE UPPER MANTLE

We investigated unique harzburgite nodule (Mc-1) from the Montana Clara Island (Canary archipelago) where we for the first time discovered evidence of primary carbonates.

The investigated xenolith with the size of 8 x 6 cm consists of olivine (86%) first and second generation, orthopyroxene (11%), spinel (<1%), clinopyroxene (2,1%), carbonate (<1%) and glass (3%). Minerals of first generation (olivine, orthopyroxene, spinel) are cut by fine-grained aggregates consisting of clynopyroxene, glass, and carbonate (Fig. 1)

Carbonate in the investigated nodules forms usually small 10-30 μm segregations of oval, rounded and wormicular shape, which are found only in association with glass, second generation olivine and spinel and also with clinopyroxene. Carbonate is enriched in calcium and has low concentrations of magnesium and iron (Table. 1). Tiny crystals of fluorite were found in carbonate. Apatite is possibly also present. There is mineral phase of rounded shape in one of carbonate segregations whose size is smaller than 1 micron that is the size of electron beam which contains chromium, cerium and lanthanum, Ce being prevailing over La. From these semiquantitative analysis one may assume that this phase represents the mixture of very fine crystals of carbonate of cerium, lanthanum and chromite or it represents mineral belonging to spinel group with the formula $AM_{12}O_{19}$, where A is Ce-La and M is Cr (krishtonite?).

Glass forms wormicular interstitial veinlets and embayments between crystals. The color of glass is slightly brownish and sometimes brown. Very often glass is associated in veinlets with rather small crystals of clinopyroxene, second generation olivine, spinel and carbonates. Glass composition plots into trachite (syenite) field with rather high agpaitic index (0.8-0.9). Very small (10 μm) globules of originally sulphide rich material are present in the silicate glass, usually very close to glass-carbonate boundaries. Microprobe analyses indicate that these globules are compositionally and mineralogically complex, with 1μm thick oxide rims and cores containing mixed sulphides and fine-grained oxide phases of Fe, Ni and Cu, including Fe-bearing bunsenite (NiO), trevorite (ideally $NiFe_2O_4$), and magnetite. Pyroxene with up to 7.5 wt% NiO has also been found. Similar oxide-sulphide mixed phases have been reported from an ophiolite complex (Tredoux et al. 1989). This phase assemblage, by analogy with sulphides in other upper mantle nodules (Lorand and Conquere, 1983), could have initially formed as immiscible Fe-, Ni-, Cu-rich monosulphide melt globules, which subsequently exsolved into pentlandite, pyrrhotite and chalcopyrite, and ultimately were altered by low temperature, oxidizing fluids.

In one thin section of the investigated xenolith we found segregation of carbonate in association with glass of rather large size 250 x 150 microns situated at the boundary between two olivine crystals (Fig.1). Carbonate is immersed into glass and forms with it distinct meniscus. Smaller segregations of carbonate form rounded or oval blebs

Table 1 Compositions of nodules and minerals and carbonates from nodule MC-1 (- not detected. NA not analyzed). Nodule analyzed by XRF at Institute of Geochemistry. Moscow: Minerals by electron microprobe at Manchester University. and Moscow With beam current 15 nA and spot size 2 μm except for glass and carbonate where beam was 10 μm.

	Nodule MC-12	Nodule MC-12	Olivine (I)ab	Spinal (I)d	Spinal (I)d	Spinal (I)d	Glass c	Carbonate c	Carbonate c	Carbonate c	Cpx (host lava)b	Ortho-pyroxene (I)d	Olivine (II)c	Clino-pyroxene (II)b	Clino-pyroxene (II)c
SiO_2	44.50	44.10	41.60	-	-	-	63.80	4.10	1.10	-	49.50	57.40	39.30	54.30	53.50
TiO_2	0.03	0.05	-	0.03	-	0.10	0.70	-	-	-	1.60	-	-	-	0.05
Al_2O_3	0.88	1.40	-	30.30	25.70	19.30	16.00	1.20	0.27	-	5.70	1.00	-	3.00	2.60
FeO	7.70	7.90	8.60	15.40	13.90	11.40	2.00	0.59	0.28	0.10	5.80	5.20	7.30	2.60	2.30
MnO	0.13	0.15	-	-	-	-	-	0.09	-	-	-	0.11	0.14	0.12	0.10
Cr_2O_3	0.30	0.57	-	39.50	44.70	53.80	0.06	0.14	0.18	-	0.57	0.50	-	0.97	0.34
NiO	NA	NA	0.31	0.10	0.10	-	-	-	-	-	-	0.11	0.37	-	0.11
MgO	45.60	44.40	50.10	15.40	15.00	15.70	2.50	6.80	3.10	3.10	14.30	34.60	51.10	17.90	18.30
CaO	0.75	0.73	0.07	-	-	-	3.80	54.30	50.80	47.00	22.80	0.21	0.20	19.80	21.50
Na_2O	0.11	0.10	-	-	-	-	6.40	0.20	0.31	0.13	0.67	-	-	1.10	0.19
K_2O	0.04	0.09	-	-	-	-	3.80	0.13	-	-	-	-	-	-	-
P_2O_5	0.05	0.05	-	-	-	-	-	1.90	-	0.15	-	-	-	-	-
S	NA	NA	-	-	-	-	-	0.24	-	0.14	-	-	-	-	-
Sum	100.1	99.5	100.7	100.7	99.4	100.3	99.1	69.7	56.0	50.6	100.9	99.1	98.4	99.8	99.0
Mg#	0.91	0.91	0.91	0.67	0.66	0.71	0.69	0.95	0.93	0.98	0.82	0.92	0.93	0.93	0.94

Table 2 Compositions of phonolite, minerals from globules and groundmass.

Elem wt.%	Phonolite	Albite Phen.	Albite G.m.	Albite Globule	Mica Phen.	Mica G.m.	Mica Globule	Calcite G.m.	Calcite Globule	Ilmenite Phen.	Ilmenite Globule	Sr-lyeshite Globule	Ancylite Globule	Apatite Globule	Strontionite Globule
SiO$_2$	47.02	67.98	68.08	68.89	37.09	37.23	36.65	0.41	0.23	0.63					
TiO$_2$	0.6				3.12	2.36	2.26	0.13		52.28	53.79	3.45	0.1		0.3
Al$_2$O$_3$	16.31	19.65	19.33	19.52	13.27	11.77	12.85	0.23	0.1	0.35		1.29		0.23	
FeO	5.01	0.19	0.10		20.17	21.50	22.18	3.14	2.42	45.43	45.14				
MnO	0.16		0.09					1.01	1.26	1.15	1.39	0.26			
MgO	2.71				13.28	11.86	12.61	0.83	0.73	0.17				0.22	
CaO	8.43	0.39		0.18	0.09	0.26	0.02	52.87	54.96	0.19		12.09	2.83	52.9	8.18
Na$_2$O	6.88	10.52	11.22	11.73	1.17	0.60	0.63	0.17	0.22	0.28	0.28	7.85			
K$_2$O	4.17	1.01	0.2	0.09	8.59	9.36	9.69	0.23							
P$_2$O$_5$	1.33													40.84	
CO$_2$	6.95														
F														3.87	
Total	100.0	99.75	99.02	100.4	96.79	95.07	94.81	60.36	61.05	100.4	100.6	93.87	75.79	98.06	68.62
SrO								1.16	0.91			5.03	27.23	1.19	58.56
BaO								0.19	0.2			0.08	0.28		1.85
Nb$_2$O$_5$												63.39			
La$_2$O$_3$												0.17	17.04		
CeO$_2$												0.24	22.47		
Nd$_2$O$_3$													5.40		

Globle-Compositions of minerals from globules.
G.m. - Compositions of minerals from ground mass
Phen. - Compositions of phenocrysts.

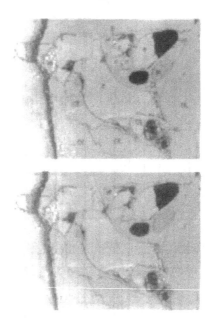

Fig. 1 Back-scattered electron images of carbonate-silicate segregations in Montana Clara nodule MC-1:Interstitial patch between first generation olivine crystals (abbreviations: *C* carbonate, *pc*-'plucked-out' carbonate, *G*-silicate glass, *S*-sulphide global, *as*-altered sulphide bleb, *Ol* -olivine)

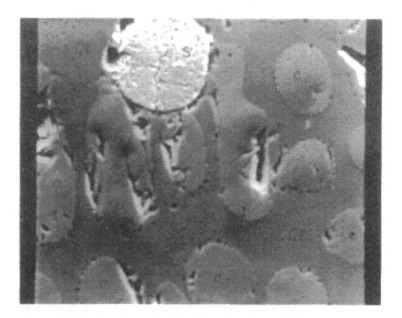

Fig. 2. Back-scattered electron image of the run product, T-1220⁰ C, P-8kb. Sulphide-silicate-carbonate immiscibility, S-sulphide liquid, C-carbonate liquid, Gl-glass, silicate liquid. Magnification ×30.000

surrounded by glass and there are holes of oval shape which obviously represent carbonate crushed out during polishing. This is high calcium carbonate with low magnesium concentration, carbonate of this composition is similar to calcite carbonatite-sovites. Textural features of the investigated nodule and chemical composition of its minerals testifyes to metasomatic interaction of harzburgite in the mantle of Montana Clara Island with primary dolomitic melt. The latter was probably transported from the lower zones of carbonatized mantle during its partial melting. As a result of this interaction following the reactions:

$$2MgSiO_3 + CaMg(CO_3)_2 = 2Mg_2SiO_4 + CaMgSi_2O_6 + 2CO_2$$
$$3CaMg(CO_3)_2 + CaMgSi_2O_6 = 4CaCO_3 + 2Mg_2SiO_4 + CO_2$$

the assemblage of secondary minerals olivine plus pyroxene, plus spinel appeared which means that partial wehrlitization of starting harzburgite took place. During this process depleted harzburgite was considerably enriched in light rare earths (Table 1) Primary carbonate melt of the mantle material of Montana Clara Island was to a large extent enriched in sodium because clinopyroxene in the investigated nodule contains up to 1.1% Na_2O which according to our calculations based on experimental data of Dalton and Wood, (1993) corresponds to 5% Na_2CO_3. Therefore metasomatic assemblage of secondary minerals was in equilibrium with sodium containing dolomitic composition. However the investigated carbonate from the nodule contains low sodium concentrations (0.1-0.3%). It is probable that sodium was dissolved by the late low temperature fluids whose evidence is manifested in oxidation and decomposition of sulfides. From the experimental data (Dernov-Pegarev, Malinin. 1976) one may conclude that nyerereite which is the main sodium compound in natrocarbonatites is dissolved incongruently under the conditions of low temp3erature hydrothermal process.Sodium carbonate enters solution during nyerereite interaction with hydrothermal fluid while solubility of calcium carbonate is lower by order of magnitude. During the reaction of primary carbonate melt with harzburgite the Ca/Mg and Mg/Fe ratios in metasomatising liquid changed. According to experimental data (Dalton and Wood, 1993) with the decreasing pressures Ca-Mg partitioning between melt and clinopyroxene shifts to more Ca rich liquid, the Ca/Ca+Mg+Fe+Na ration in carbonate melt reach up to 0.92. Magnesium to iron ratio also increases. Calcium-rich carbonate confined to wehrlite veinlets of the investigated nodule and characterised by very high Ca/Ca+Mg+Fe+Na (0.84-0.91) and Mg/MgtFe-(0.95) ratios is the result of reactions taking place at pressures below 20 kb.

Using two pyroxene thermometers (Brey, Kohler, 1990) we estimated lower temperature limit of metasomatic reaction between carbonate melt and harzburgite because in this case orthopyroxene isan unstable phase in this process. The estimated temperatures fall into range the 1150-1075° C. Oxygen fugacity was estimated from olivine-spinel-orthopyroxene geobarometer (Wood et al.. 1990) and gave a range of 0.76-to 0.06 log units below the QFM (quartz-fayalite-magnetite) buffer.

Our investigations revealed very complicated geological history of the mantle material of Montana Clara Island. The extremely depleted character of the investigated xenolith (Table 1) points to the processes of substantial - more than 25% melting during which the initial substrate lost so-called basaltic elements: CaO. Al_2O_3 .TiO_2, Na_2O and others. All the sulfur should have been lost during this process due to the low melting temperatures of mantle sulphides. Later this depleted mantle material was penetrated by metasomatising carbonate melt with the prevailing dolomite component and containing sodium and possibly potassium, sulfur, light rare earths, strontium and phosphorus. During the metasomatic reactions partial wehrlitisation along the veinlets and fissures of the initial harzburgite took place. Later, during the very rapid ascend of mantle material (otherwise primary carbonate would not survive) wehrlitic mineral assemblage which contained carbonate melted during decompression. The character of interrelationships between carbonate, sulphide globules and glass (Fig.1) testifies to the processes of immiscibility between carbonate, silicate and sulphide melts.

The immiscibility between silicate and sulphide melts was known for many years. Many authors also reported immiscibility between silicate and carbonate liquids in a wide range of temperatures and pressures (Koster van Groos, Wyllie, 1966 and many others). However, the immiscibility in carbonate-silicate- sulphide systems has so far not been revealed. We have investigated using the piston-cylinder apparatus the immiscibility in the system Ca-rich carbonate- Fe,Ni sulphide-silicate melt of phonolitic composition containing F. Experiments were made at 1250°C and 4-15 kbar. Double Pt capsule method has been employed in order to control oxygen fugacity. The immiscibility has been observed in the investigated system exhibited in the complete separation of carbonate and silicate liquids, whereas sulphide melt was present in the form of small globules in both liquids (Fig.2). Sulphur solubility in silicate melt varies from 0.15 to 0.35% and in carbonate liquid it ranges from 0.02 to 3.7% depending on alkali content. These results permit to suggest that the immiscibility of carbonate, sulphide and silicate melts took place during the partial melting of upper mantle material. The glass in the investigated nodule is enriched in alkalis and silica probably due to incongruent melting of jadeite-bearing clinopyroxene during the reduction of pressure.

In summary the investigated mineral assemblage carbonate + glass + sulfides + clinopyroxene + orthopyroxene + spinel is the result of the process of metasomatic interaction of depleted mantle with carbonate melt and subsequent partial melting resulted in the formation of carbonate liquids similar to carbonatites. It may be concluded that high-calcium carbonate melts may be formed during partial melting of wehrlitic mantle. It may be noted in this respect that earlier (Ryabchikov et al., 1989) we have found xenoliths of carbonated wehrlite with calcium-rich carbonate containing 3.52% MgO on San-Vincente Island (Cape Verde Islands) where carbonatite volcanism is wide spread.

Therefore the investigated unique mineral assemblage including carbonate and glass can be considered as a micro model of the generation of the carbonatitic magmas during the processes of the partial melting of wehrlitized and carbonatized mantle. The development of carbonatite magmatism on Canary Islands, Cape Verde Islands is likely to be related to the partial melting of carbonatized mantle in the South Atlantic, which took place over a vast territory.

THE ROLE OF CO_2 IN DIFFERENTIATION OF ULTRAMAFIC ALKALINE SERIES. LIQUID IMMISCIBILITY IN CARBONATE BEARING PHONOLITIC DYKES (POLAR SIBERIA)

One of the largest of its kind in the world, Maimeicha-Kotui province of ultramafic alkaline rocks, is situated in the North of Siberian platform in Polar Siberia and extends over an area 220 x 350 km between Maimecha and Kotui rivers.(Egorov, 1991; Kogarko et al., 1995). There are 37 ultramafic alkaline and carbonatitic massifs of complicated structure. This province is particularly noteworthy not only for the abundance of carbonatites but for the association of many of them with ultramafic rocks. The representive rocks types include dunite, pyroxenite, ijolite, melteigite, jacupirangite, phoscorite, a range of melilite-bearing rocks and numerous and varied carbonatites. There are hundreds of dykes, including radial dykes which surround the ultramafic-alkaline massifs (for instance Bor-juryakh, Romanicha, Odichincha, Dolbykha, Kugda)

We investigated a series of radial dykes of Dolbykha massif which comprise alnoite, olivine-nephelinites, nephelinites, melilite nephelinites, cancrinite, nosean and calcite phonolites, calcite trachytes, calcite carbonatites and monchiquites.

Some dyke phonolites contain carbonate-bearing globules of rounded and oval shapes with the sizes of 1-2 mm to 17-20 mm. One globule-bearing phonolite (N 873) was chosen for detailed investigation. Phonolite consists of phenocrysts (~10%) represented by albite, mica, and ilmenite, globules (~ 20%) and groundmass (~70 %). Globules consist of polycrystalline calcite aggregate and contain albite, mica, apatite, cancrinite, Sr-lueshite, zircon, ancylite, ilmenite, strontianite (Fig.3, 4). Occasionally the reaction relations between mica and albite are observed. Albite grains form rims around mica. Carbonate- bearing globules are characterised by sharp contact with ground mass, and they are seemingly flowed around by the crystals of groundmass (mainly albite).Groundmass is fine-grained and it includes albite and mica with subordinate K-feldspar and nepheline. Among accessory minerals apatite, ilmenite, ancylite, and Sr-lueshite were obsereved. Representative microprobe analyses of mineral phases are given in Table 2. The substantial substitution of Ca by Sr in apatite and calcite is observed. Similar feature was described by Kjarsgaard and Peterson (1991) in globular nephelinite from Shombole.

The consideration of this material in the light of experimental data on the liquid immiscibility in carbonate-silicate systems suggests that the separation of carbonatite melts from phonolitic ones took place due to the immiscibility in liquid state.

Petrographic data support this hypothesis because carbonate globules are obviously flown around by the crystals of ground mass (albite and mica). It may be noted that there is the great similarity in composition of minerals in groundmass and in globules, which demonstrates the existence of equilibrium in this carbonate-bearing phonolitic system (Table 2). As has been already noted the increasing alkalinity of carbonate-bearing magmatic systems favours the appearance of immiscible liquids. It should be pointed out that the boundary of liquid immiscibility field in the system $Na_2O+K_2O-Al_2O_3+SiO_2-CaO-CO_2$ corresponds to agpaicity index Na_2O+K_2O/Al_2O_3 close to 1. (see Fig 15-1, Bell, 1989). This implies that in agpaitic melts with liquid immiscibility is more likely to occur. The investigated phonolite belongs to peralkaline type (agp. ind. = 1.1), and therefore formation of immiscible liquids is more probable in it . The crystallisation of mica phenocryst enriched in Al also results in the increase in peralkalinity of residual melt.

The presence of Sr lueshite in globules of investigated phonolite also supports the oversaturated in respect alkalis character of carbonate melt because lueshite can crystallize only from peralkaline compositions according our experimental data on $Na_2O+K_2O-SiO_2+Al_2O_3-CaO-CO_2$ system in which the direction of tie-line corresponds   the more alkaline composition of equilibrium carbonate immiscible liquids. The alkalis were subsequently lost into fluid phase due to incongruent (Kogarko et al., 1982). We propose that originally carbonate melts contained significantly higher alkali amounts.

L. N. Kogarko

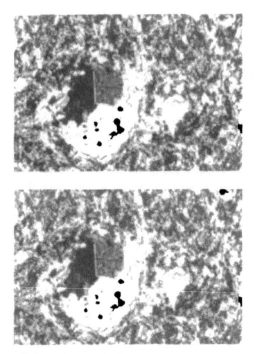

Fig. 3 Carbonate globule in phonolite N 873 crossed nicoles magnification × 250

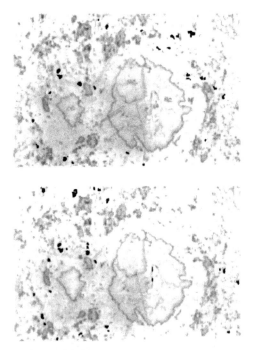

Fig. 4 The same carbonate globule in phonolite N 873, reflected light mc – mica, Anc – ancylite, Ab – albite, cc – calcite, Ap – apatite

on $Na_2O+K_2O-SiO_2+Al_2O_3-CaO-CO_2$ system in which the direction of tie-line corresponds the more alkaline composition of equilibrium carbonate immiscible liquids. The alkalis were subsequently lost into fluid phase due to incongruent (Kogarko et al., 1982) dissolution of calcium-sodium carbonate in aqueous fluid at lower temperatures (Dernov-Pegarev V.F., Malinin S.D., 1976). Our discovery of nyererite as solid microinclusions in perovskite in the Guli carbonatite of Polar Siberia (Kogarko et al, 1991) confirms this assumption. From the recent experimental data of Woh-jer and Wyllie (1995) we can conclude that under the higher pressure (2.5 GPa) liquid immiscibility does exist even in miaskitic melts while under lower pressure (1 Gpa) in the same system immiscibility phenomena is absent mostly due to the degassing. Calcite, melilite and wollastonite may crystallize as liquidus phases if the melt becomes very Ca-rich. In one phonolitic dyke (N 2003) of Dolbykha complex there is liquidus calcite as the phenocrysts without any sign of immiscibility phenomenon.(Fig.5).

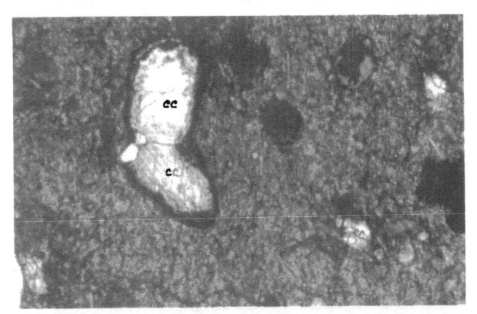

Fig. 5 Phenocrysts of calcite in phonolite N 2003, mg 100 crossed nicols. Magnification ×250.

The regime of CO_2 in alkaline systems controls differentiation of residual phonolitic magmas. Under high pressure of carbon dioxide the separation of an immiscibility carbonate liquid occurs. At lower P_{CO2} values liquid immiscibility is not observed and calcite may crystallize either as phenocryst or in groundmass. At still lower pressure carbonates are not formed due to the loss of carbon dioxide and mineral assemblages including either melilite or wollastonate occur. Some melilite-bearing rocks of Polar Siberia are result of this process.

Thus one of the important mechanisms of the genesis of carbonatite melts in Polar Siberia were phenomena of liquid immiscibility in strongly differentiated phonolitic magmas. The generation of the carbonatites is probably controlled by the depth (and

P $_{CO_2}$) of intermediate magma chamber where differentiation took place, and probably by the alkalinity of melts, and speed by the of magma ascend to the surface.

CONCLUSION

The above examples demonstrate an important role of liquid immiscibility in the genesis of Ca-rich carbonatite magmas. Formation of immiscible liquids during processes proceeding in upper mantle results in the generation of phonolite and carbonatite liquids.

Among the continental carbonatite formations there are Ca-rich carbonatites exclusively occuring with syenites Okorusu, Namibia (Prins, 1981), Siilinjarvi, Finland (Puustinen 1969), Stjernoy, Norway (Robins 1971), and Vishnevye Gory, Russia (Kogarko et al. 1995). It is possible to suggest the origin involves liquid immiscibility in these cases. It should be pointed out that the Phalabora Complex, South Africa, which contains satellite plug of alkali syenite, alkali quartz syenite and alkali granite in association with calciocarbonatite (Eriksson,1989), is also characterized by the presence of Cu-Fe sulphide deposits. This leads us to speculate that calciocarbonatite-silicate-sulphide immiscibility similar to that observed in the Montana Clara nodule might have played some role in genesis of this complex.

Under the conditions of earth's crust the carbonate-silicate liquid immiscibility arises at the late stages during the prolonged differentiation of ultrabasic-alkaline magma. The principal factors controlling the generation of carbonatites during formation of immiscible liquids are CO_2 partial pressure and alkalinity of magmatic systems.

Acknoledgments

This work was done with financial support of RFBR N 96-05-64151 and INTAS 95-IN-RU-953

REFERENCES

1. H.E.F. Amundsen. Evidence for liquid immiscibility in the upper mantle. *Nature* 327, 692- 696 (1987).
2. M.B.Baker, P.J. Wyllie. Liquid immiscibility in a nephelinite-carbonate system at 25 kbar and implications for carbonatite origin, *Nature* 346, 168-170 (1990).
3. D.K.Bailey. Carbonate melt from the mantle in the volcanoes of south-east Zambia. *Nature* 338, 415-418 (1989).
4. K.Bell (ed.). Carbonatites: Genesis and Evolution. Unwin Hyman, London, 618 pp. (1989).
5. G. Brey, T.Kohler. Geothermometry in four-phase lherzolities. II. New thermobarometers and practical assessment of existing thermobarometries. *Geochim Cosmochim Acta* 31: 1353 -1378 (1990).
6. G.A.Dalton, B.J.Wood. The compositions of primary carbonate melts and their evolution through wallrock reactions in the mantle. *Earth Planet Sci Lett* 119: 511-525 (1993).
7. V.F.Dernov-Pegarev, S.D.Malinin. Calcite solubility in high temperature aqueous solutions of alkaline carbonatites and problems of carbonatite formation. *Geochemie* 5: 643-657 (1976).

8. D.H. Eggler. Carbonatites. primary melts, and mantle dynamics, in K. Bell (ed.), Carbonatites Genesis and Evolution. Unwin Hyman, London, pp. 561-579 (1989).

9. L.S.Egorov. Ijolite carbonatite plutonism (case history of the Maimecha-Kotuy complexes of northern Siberia) Nedra, Leningrad, 260 pp.(1991).

10. S.C.Eriksson. Phalaborwa: a saga of magmatism, metasomatism, and miscibility. In: Bell K (ed). Carbonatites genesis and evolution. Unwin Hyman, London, pp 221-249 (1989).

11. I.C.Freestone & D.L. Hamilton. The role of liquid immiscibility in the genesis of carbonatites. *Contributions to Mineralogy and Petrology* 73 105-117 (1980).

12. D.L. Hamilton., P. Bedson, J. Esson. The behaviour of trace elements in the evolution of carbonatites, in K. Bell (ed)., Carbonatites: Genesis and Evolution. Unvin Hyman, 405-427 (1989).

13. B.A.Kjarsgaard and T.D. Peterson. Nethelinite-carbonatite liquid immiscibility at Shombole Volcano, East Africa *Mineralogy and Petrology* 43, 293-314 (1991).

14. L.N.Kogarko, A.A.Tugarinov, L.D. Krigman. Phase equilibria in the nepheline-luestite system. *Doclady Academii Nauk* 263, 985-987 (1982)

15. L.N. Kogarko, D.A.Plant, C.M.B.Henderson and B.A Kjarsgaard. Na rich carbonate inclussion in perovskite and calzirtite from the Guli intrusive Ca-carbonatite, Polar Siberia. *Contributions to Mineralogy and Petrology* 109, 124-129 (1991).

16. L.N Kogarko. Geochemical characteristics of oceanic carbonatites from the Cape Verde Islands. *S.Afr.Geol.* 96 ,119-125 (1993).

17. L.N.Kogarko, V.A. Kononova, M.P. Orlova and A.R.Woolley. Alkaline Rocks and Carbonatites of the World. Part 2: Former Soviet Union

18. I. Henderson., A.H. Pacheco. *Contribution to Mineralogy and Petrology.* 121, 267-275. (1995b).

19. A.F.Koster van Groos and P.J.Wyllie. Liquid immiscibility in the system Na_2O-Al_2O_3-SiO_2-CO_2, *Amer. Joural Sci.* 264, 234-255 (1966).

20. A.F.Koster van Groos and P.J.Wyllie. Liquid immiscibility in the join $NaAlSi_3O_8$-Na_2CO_3-H_2O and its bearing on the genesis of carbonatites, *Amer. Joural Sci.* 266, 932-967 (1968)

21. J.P.Lorand, F. Conquere. Contribution a l'etude des sulfures dans les enclaves de lherzolite a spinelle des basaltes alcalins (Massive Central et Languedoc, France). *Bull Mineral* 106:585-605

22. T.D.Peterson. Peralkaline nephelinites I. Comparattive petrology of Shombole and Oldoinyo. *Contrtibution to Mineralogy and Petrology.* 101, 458-478 (1989).

23. P.Prins. The geochemical evolution of the alkaline and carbonatite complexes of the Damaraland Igneous Province. *S.W. Africa Ann Univ Stellenbosch Ser Al Geol* 3 : 145-278 (1981).

24. K. Puustinen. The carbonatite of Siilinjarvi in the Precambrian of eastern Finland: a preliminary report , *Lithos* 3 : 89-92 (1969).

25. J.M.Pyle & S.E. Haggerty. Silicate-carbonate liquid immiscibility in upper-mantle eclogites: implications for natrosilicic and carbonatitic conjugate melts. *Geochimica et Cosmochimica Acta* .58, 2997-3011 (1994).

26. B.Robins. Syenite-carbonatite relationships in the Seiland Gabbro Province, Finnmark, northern Norway, *Nor Geol Unders* 272: 43-58 (1971).

27. I.D. Ryabchikov, G.Brey, L.N.Kogarko, V.K. Bulatov. Partial melting of carbonated peridotite at 50 kb. *Geokhimia* 1:3-9 (1989)

28. M. Tredoux, M.J.de Wit, R.J.Hart, R.A.Armstrong, N.N.M.Linsay. J.P.F. Sellschop. Platinum group elements in a 3.5 Ga nickel-iron occurrence: possible evidence of deep mantle origin, *J Geophys Res* 94B: 795-813 (1989).

29. J.D.Twyman and J.Gittins in J.G. Fitton and B.G.J. Upton (eds). Alkaline igneous rocks, Oxford and London, Blackwell Scientific, Geological Society Special Publication No.30, 85-94 (1987).

30. L. Woh-jer and P. J.Wyllie. Experimental data bearihg on liquid immiscibility, crystal fractionation, and the origin of calciocarbonatites and natrocarbonatites. *International Geological Review*. 36, 797-819 (1994).

31. L.Woh-jer and P.J. Wyllie. Liquid immiscibility and calciocarbonatite magmas. *Submitted to Journal of Petrology*. (1995).

32. B.J.Wood, L.T.Bryndzya, K.E.Johnsen. Mantle oxidation state and its relationship to tectonic environment and fluid speciation. *Science* 248:337-345 (1990).

33. P.J Wyllie. Origin of carbonatites: evidence from phase equilibrium studies. In K. Bell.Carbonatites: Genesis and Evolution: London. Unwin Hyman. London,500-540 (1989).

34. P.J.Wyllie, M.B.Baker and B.S. White. Experiment boundaries for the origin and evolution of carbonatites . *Lithos*. 26,3-19 (1990).

Proc. 30ᵗʰ Int'l. Geol. Congr., Part 15, pp. 65 – 86
Li et al. (Eds)

China's Igneous Rock Associations and Their Temporal–Spatial Distribution and Relations to the Structure and Composition of the Crust and Mantle

LI ZHAONAI and WANG BIXIANG

Institute of Geology, Chinese Academy of Geological Sciences, Beijing, 100037 CHINA

Abstract

The Geological Map of Igneous Rocks of China (1:5 000 000) compiled by Li Zhaonai, Wang Bixiang and other editors portrays China's geological settings, embracing tectono–magmatic domains; major faults, volcano centers, ages of volcanic rocks and plutonic rocks and Moho depth contours, with emphasis on associations of volcanic and plutonic rocks. In the classification of igneous rock associations the editors consider their geological environments and geological, petrological and mineralogical features as well as their major and trace element and isotope geochemical features. In the map essential rock associations are expressed and the Early Proterozoic to Quaternary ages of volcanic and plutonic rocks are marked. The contents of this map indicates the following: the Chinese continent originated by assembly of multiple blocks and different blocks have brought about the three–dimensional inhomogeneity of the structure and composition of the crust and mantle of the continent; the features of various volcanic and plutonic rock associations imply that they occur respectively in different second–order geological environments of three kinds of tectonic settings: collisional orogenic belts, stable continental blocks and continental activation belts; trace element and isotope geochemical features of different magma types suggest that they are the products of interactions of different blocks and different crustal and mantle layers.

Key words: igneous rock association, China

INTRODUCTION

The Geological Map of Igneous Rocks of China (1:5 000 000) was produced as one of the projects of map series compilation assigned by the Ministry of Geology and Mineral Resources (MGMR) of the People's Republic of China.

The map portrays such essential elements as the tectono – magmatic domains, major fracture systems, ages of volcanic rocks, volcano centers and Moho depth, with emphasis on the natural associations of volcanic and plutonic rocks. In the classification of igneous rock associations, we consider the geological setting, geology, petrology, mineralogy, major and trace element geochemistry and isotope geochemistry. In the map 27 associations of volcanic and plutonic rocks are distinguished and 54 major rock types are portrayed. The age range is from the Proterozoic to Quaternary, and the smallest geochronologic unit indicated in the map is epoch.

The complex volcanic and plutonic rock associations of different ages, from an angle of magmatism, reflect the three – dimensional inhomogeneity of the crust – mantle structure and composition of the Chinese continent formed by assembly of multiple blocks, the complexity of the process of interactions of different blocks and layers in the interior of the continent and the interactions of the Chinese continent with its surrounding plates.

GEOLOGICAL SETTINGS OF IGNEOUS ROCKS

The present Chinese continent was formed by assembly of many blocks and each block had its own process of development. The blocks that have a relatively complete history of evolution has largely gone through the following stages: the stage of continental nucleus and block development prior to the early Late Proterozoic, the stage of continental – margin development from the late Late Proterozoic to early Mesozoic, and the stage of intracontinental development from the Middle Mesozoic to Cenozoic. As the Chinese continent as a whole is concerned, the developments are unbalanced. The evolutions of different blocks kept step with each other but did not occur entirely synchronously.

Regionalization of Geological Settings

On the basis of the crust – mantle structure, composition and plate – tectonic regionalization, six major regions of geological settings of igneous rocks in China may be distinguished. From north to south and from west to east they are: the Junggar – Hinggan domain (I), Tarim – North China domain (II), Hoh Xil – South China domain (III), Zangdian (Tibet – Yunnan) domain (IV), Himalaya domain (V) and East Taiwan domain (VI). (1) The Junggar – Hinggan domain is equivalent to the Paleozoic composite orogenic belt on the southern margin of the Siberian plate, consisting of the Altay – Ertix and Junggar – Hinggan epicontinental accretionary fold belts (including the Junggar, Xilinhot, Nenjiang and Jiamusi massifs) and the Wanda slab. (2) The Tarim – North China domain is in the main equivalent to the Tarim – North China plate, composed of the Tianshan – Chifeng fold system on the northern margin and the West Kunlun – Altun and North Qilian – North Qinling rift fold belts (including the Qaidam microblock) on the southern margin. (3) The Hoh Xil – South China domain is generally consistent with the scope of the South China plate including the northern Hubei – South Qinling and Zhangbaling – Haizhou Paleozoic – initial Mesozoic collisional fold belt on the northern margin, the Yangtze block, Nanhua (a part of the South China) Early Paleozoic fold belt, Hoh Xil – Songpan – Garzê Early Mesozoic fold belt, northern Qiangtang – Qamdo – Simao micro – block and Qiongnan (southern Hainan) micro – block. (4) Zangdian (Xizang – Yunnan) domain is equivalent to the Zangdian plate, consisting of the Gangdisê – Tengchong Meso – Cenozoic fold belt and south – central Qiangtang – Tanggula – Baoshan blocks. (5) The Himalaya domain belongs to the northern margin of the Indian plate, consisting of the Himalayan Cenozoic fold belt and the northern margin of the Indian block. (6) The East Taiwan domain belongs to the Cenozoic subduction – collision fold belt in eastern Taiwan on the western margin of the Philippine Sea plate.

Fault Systems

On the basis of their ages of formation and relations with the stages of tectonic development, faults and great faults of the Chinese continent may fall into four basic tectonic systems, i.e. the Paleo – Asiatic system, Tethys system, Paleocathaysian – peri – Pacific system and Helan – Kangdian (Xikang – Yunnan) system. (1) The faults of the Paleo – Asiatic system are mainly distributed in the Junggar – Hinggan domain, the Tarim – North China domain and the fold belt on the northern margin of the South China domain. They are the product of interactions of the Siberian plate, Tarim – North China plate and South China plate during their repeated convergence and divergence since the Paleozoic. They, distributed mainly in an E – W direction, may be divided into the north and south fault systems. The north fault system includes the Kelamaili – Erenhot, Ertix – Derbugan and North China northern – margin fault belts. These fault belts strike E – W in the central part, and turn WNW in the western part and NE and NNE in the eastern part. The south fault system is called the Kunlun – Qinling fault system, distributed generally in a WNW direction. With the Kunlun – Qinling junction zone as the trunk, it comprises the Kunlun – Qinling composite fold system and the thrust nappe zone on the northern margin of the Yangtze block. The derivative faults associated with the trunk fault include nearly N – S – trending tensional faults and ENE – NNE and WNW – NNW conjugate shear faults; besides, the large – scale Altun strike – slip fault system in the southeastern part of the Tarim block obviously show a sinistral strike – slip nature. (2) The faults of the Tethys system are dominantly distributed in the Zangdian plate and some lie on the western and southwestern margins of the South China plate, including the Paleo – Tethys, Neo – Tethys and Youjiang River fault systems. They resulted from recurrent over – all or local collision of the Zangdian plate with the Yangtze and Indian plates and their extension from the Late Paleozoic to Meso – Cenozoic. Among other things, the Paleo – Tethys fault system is made up of the Lancang River junction zone and the Jinshan River and Nujiang River crustal assembly zones. The fault system strikes NW in the northern part, nearly N – S in the central part and SE in the southern part, showing a dextral strike – slip nature. The Neo – Tethys fault system with the Yarlung Zangbo junction zone as the trunk strikes nearly E – W, with its eastern sector striking nearly N – S. The Himalaya region is a strong thrust zone. The Youjiang fault system with the Ziyun – Nandan and Youjiang faults as the trunks strikes NW, showing a sinister strike – slip nature. (3) Except those old faults involved into the Sibao'an Shexian – Dexing and Shaoxing – Pingxiang – Beihai assembly zones, the faults of the Paleocathaysian – peri – Pacific system may be broadly divided into the NE – and NNE – trending fault systems, of which the NE – trending fault system is mainly distributed in the coastal area of southeastern China and dominated by thrust faults, including the Nan'ao – Honggang (Xianggang), Lishui – Lianhuashan, Chong'an – Heyuan, Fuzhou – Suichuan and Changshoujie – Shuangpai faults. The NNE – trending fault system, which has a sinistral strike – slip nature, is mainly distributed in the peri – Pacific region, comprises the Da Hinggan – Wuling, Tancheng – Lujiang and Taiwan faults. (4) The Helan – Kangdian fault system in the main strikes nearly N – S. The faults occurring in the Panxi (western Panzhihua) area include the Lüzhi River, Anning River and Xiaojiang River faults. They originated in the Proterozoic and then were compounded into a fault system passing from north to south through the Chinese continent in the Meso–Cenozoic as a result Tethyan and peri–Pacific tectonism.

ASSOCIATIONS OF VOLCANIC ROCKS

The nomenclature of igneous rocks here is in principle based on the classification schemes of volcanic and plutonic rocks recommended by the IUGS Subcommission on the Systematics of Igneous Rocks (Streckeisen, 1978; Le Bas et al., 1986), while the nomenclature of the rocks that are not concerned by the IUGS is based on the classification schemes of pyroclastic rocks and lavas recommended by the Commission on Petrology of the Geological Society of China.

The basic principles for the classification of natural associations of volcanic and plutonic rocks are as follows. (1) A group of specific varieties of volcanic or plutonic rocks are closely related to one another by origin. (2) Different rock types of the same association have gradational relationships or show certain similarities and correlation in respect of petrology, mineralogy and element and isotope geochemistry. (3) The same rock associations occur in similar geological settings. (4) Different rock types in the same rock association originate in the same time range and is the product of the same (or the same group of) magmatic processes. (5) The rock associations have rather complex origins; on the basis of the origins, they may broadly fall into three types of association: comagmatic differnetiated type association, mixed — magmatic differentiated type and differental magmatic syn — emplacement type.

Association consisting dominantly of acid volcanic rocks

Geological setting   (1) Intermediate — acid and acid volcanic rocks occurring in island — arc and active continental — margin environments in subduction and collision orogenic belts are mainly represented by calc — alkaline dacitic and rhyolitic volcanic rocks; (2) volcanic rocks of the same compositions as above occurring in superimposed downfaulted basins of continental activated belts and some occurring in the late island arc or post — island arc stage consist dominantly of high — K dacitic and rhyolitic or trachydacitic and rhyolitic rocks.

Association types of volcanic rocks   The associations of intermediate — acid and acid volcanic rocks may be divided into five basic types: (1) association of dacitic — rhyolitic rocks with large amounts of ordinary sedimentary rocks; (2) association of dacitic — rhyolitic rocks with varying amounts of andesitic rocks and minor amounts of basaltic rocks; (3) association of high — K dacitic — rhyolitic or trachydacitic — rhyolitic rocks with some trachyandesitic rocks and minor tranchybasaltic rocks; (4) association of panteller — alkali rhyolitic rocks with alkali quartz trachytic rocks; (5) bimodal association of calc — alkali, alkali — calcic or alkali dacitic and rhyolitic rocks and basaltic or basalt — andesitic rocks.

Textures, structures and material composition of rocks   The common feature of calc — alkaline, alkali — calcic and alkaline dacitic and rhyolitic rocks is that they are all composed dominantly of ash flow tuffs. Each cooling unit— sheet has a ternary structure, i.e. the bottom is represented by weakly welded tuff of base — surge deposits, containing exotic rigid fragments and showing oblique beding; the bulk part consists of ignimbrite with varying degrees of welding, containing varying amounts of plastically deformed lenticular pumice fragments and rigid fragments;

the top is marked by vitric tuff of fall deposits. Crystal crystals are dominated by angular quartz and alkali feldspar, with varying amounts of acid plagioclase in dacitic rocks; biotite or amphibole, commonly having opacitized rims, is rare or absent. In comparison with orogenic calc − alkaline dacitic and rhyolitic rocks, the crystal fragments of continental trachydacitic rocks are dominated by alkali feldspar, especially K − feldspar (sanidine and orthoclase), with less or no quartz; while continental rhyolitic rocks, especially the varieties with $SiO_2 > 75\%$, contain abundant quartz crystal fragments.

Geochemical characteristics of rocks   The SiO_2 content of orogenic calc − alkaline dacitic and rhyolitic rocks mainly ranges between 65% and 75%; the corresponding total alkali content is <7% to 8%, of which the K_2O content is <2.75% to 3.75%. There is a positive correlation between SiO_2 and total alkalis. For trace elements, contents of large − ion lithophile elements such as Rb, Sr and Ba are relatively high; while such transiton metal elements as Cr, Ni and Co are relatively low. The chondrite − normalized rare earth element ($\sum$ REE) content is 70×10^{-6} to 150%, with a La / Yb ratio of 6 to 16; the initial ratio of Sr isotopes is 0.7032 to 0.7043. The continental activation belt is characterized by high − K contents in dacitic rocks and high SiO_2 in rhyolitic rocks. The SiO_2 content is generally in the range of 68% to 78% and may be higher than 80% in some rhyolitic rocks. Contents of Rb, Sr and Ba are much higher and Cr, Ni and Co are lower. The chondrite − normalized $\sum$ REE content is 140×10^{-6} to 250×10^{-6}; La / Yb is 11 to 23; $(^{87}Sr / ^{86}Sr)_i$ is 0.7053 to 0.7078 and may attain 0.7095 to 0.7015 in some peraluminous varieties.

Other associated rocks   The associated rocks of dacitic and rhyolitic rocks in orogenic belts are somewhat different from those in continental activation belts: the former tend to be associated with large amounts of marine normal − sedimentary clastic rocks and clastic carbonate rocks; the latter tend to be associated with continental tuffite and tuffaceous sandstone which have similar compositions.

Association consisting dominantly of intermediate volcanic rocks

Geological setting   The calc − alkaline andesitic rock association mainly occurs in island − arc and active continental − margin environments, making up the bulk of orogenic volcanic rocks. The calc − alkaline, alkaline or alkali − calcic trachyandesitic rock association mainly occur in inherited down − faulted basins of continental activation belts and in some cases it originated in the late stage of mature island arcs or in their postmature stage and is commonly distributed in pull − apart basins near large strike − slip fault zones.

Association types of volcanic rocks   The associations of volcanic rocks mainly of intermediate composition may fall into five types: (1) andesitic lavas − pyroclastic rocks associated with large amounts of normal sedimentary rocks; (2) mainly basalt andesitic − andesitic rocks associated with varying amounts of dacitic and rhyolitic rocks and small amounts of basaltic rocks; (3) mainly sodic basalt trachyandesitic − trachyandesitic rocks associated with small amounts of andesine basaltic, quartz trachyandesitic and trachydacitic rocks; (4) mainly potassic basalt trachyandesitic − trachyandesitic rocks associated with small amounts of trachybasaltic, trachytic and phonolitic rock; (5) mainly trachytic and phololitic rocks associated with varying amounts of trachyandesitic rocks.

Textures, structures and mineral composition of rocks   Basalt andesitic and andesitic rocks mainly occur as lava flows and about 1/3 of them occur as pyroclastic rocks. At the top and bottom of the lava flow, low − Al andesite has amygdaloidal structure and the high − Al variety has brecciated lavas. The central facies has massive structure and intergranular or pilotaxitic texture. The essential minerals of orogenic calc − alkaline andesite are plagioclase (andesine on the average) and clinopyroxene with minor biotite, amphibole and orthopyroxene. The essential minerals of the sodic variety of alkali − calcic trachyandesite in continental activation belts are plagioclase (andesine or oligoclase andesine on the average) and clinopyroxene; the minor minerals are amphibole and biotite, in addition to varying amount of hypersthene. Potassic trachyandesite is characterized by the appearance of K − feldspar, which occurs as rims of independent plagioclase phenocrysts and in the interstices of matrix plagioclase microlites. Ferromagnesian minerals include clinopyroxene, biotite and minor hypersthene (which are apt to undergo secondary alteration).

Geochemical characteristics of rocks   For the major elements, the SiO_2 content of calc − alkaline basalt andesitic and andesitic rocks range from 52% to 65%, the average SiO_2 content of basaltic andesite is 54.27% and that of andesite is 59.96%. The AlK range is lower than 4.72 to 6.21% and K_2O range is lower than 0.92 to 2.16%. The average AlK of basaltic andesite is 3.93% and K_2O is 1.16%; the average AlK of andesite is 2.57. For trace elements (in ppm), Rb is 17.6 to 43.5, Sr is 154 to 186.2, Ba is 19.3 to 258, Cr is 44.8 to 258.6 and Ni is 23.5 to 97.6; $\sum REE$ is 93.5 − 133.8, La/Yb is 13 − 16 and $(^{87}Sr/^{86}Sr)_i$ is 0.7024 − 0.7043. The SiO_2 content of continental basalt − trachyandesitic and trachyandesitic rocks is 50.56% to 58.14%, with an average of 52.57% for basalt − trachyandesite and 54.57% for trachyandesite. The AlK content range is 4.81% to 6.47% for basalt − trachyandesite, 6.55% to 7.60% for sodic trachyandesite and 7.15% to 9.58% for potassic trachyandesite. The K_2O content averages $2.25 \pm 1.15\%$ for the sodic variety and $3.68 \pm 1.15\%$ for the potassic variety. Rb = 95 − 251 ppm; Sr = 211 − 720 ppm; Ba = 358 − 493 ppm; Cr = 4 − 180 ppm, Ni = 4 − 23 ppm, $\sum REE$ = 152 − 230 ppm, La/Yb = 9 − 17 and $(^{87}Sr/^{86}Sr)_i$ = 0.70524 − 0.70643.

Other associated rocks   Orogenic calc − alkaline andesitic rocks may be associated with marine or paralic secondary pyroclastic rocks and normal sedimentary rocks; trachyandesitic rocks in continental activation zones and at some postmature island arcs are commonly associated with continental tuffite, tuffaceous sandstone and terrestrial secondary pyroclastic − sedimentary rocks.

Associations consisting dominantly of basic volcanic rocks

Geological settings   Basic volcanic rocks mainly occur in three geological settings: (1) assembly zones of different plates or blocks where they occur as a part of ophiolites; (2) island arcs and active continental margins where continental crust is thin during the early stage; (3) continental rifts or back − arc pull − apart basins.

Association types of volcanic rocks   There are five major types: (1) the association of basaltic rocks and large amounts of normal sedimentary rocks; (2) the association of oceanic tholeiitic rocks with ultramafic rocks of the assembly zoe; (3) the association mainly of subalkaline basaltic rocks, including tholeiitic and calc − alkaline basaltic (low − Al or high − Al) rocks; (4) the bimodal association consisting dominantly of basaltic and basalt − andesitic rocks with subordinate rhyolitic − dacitic and rhyolitic rocks; (5) the association of continental rift volcanic rocks composed of different proportions of alkaline and tholeiitic rocks.

Textures, structures and mineral composition of rocks   Most of basaltic rocks occur as lava flow.   Their top and bottom have amygdaloidal structure, and marine basaltic rocks have pillow structure and quenched brecciated structure. The marginal facies or quenched clasts of lava commonly has vitric or semicrystalline texture, while the central facies have holocrystalline or microphaneritic texture, and intergranular, diabasic and feldspathic poikilotopic texture are common. Subalkaline basaltic rocks include olivine tholeiite, tholeiite, quartz tholeiite, high − Al basalt and low − Al calc − alkaline basalt. The essential minerals are labrador and clinopyroxene, and there may appear minor olivine in more basic varieties, minor pigeonite in the matrix of tholeiite, minor hypersthene or its standard molecules in calc − alkaline varieties, and minor quartz in silica − oversaturated tholeiite. Alkaline basaltic rocks include alkaline olivine basalt, basanite / tephrite, phonolitic basanite / tephrite, basanitic / tephritic phonolite, basic phonolite and foidite. The essential minerals include basic plagioclase, titanium − and alkali − rich clinopyroxene with varying amounts of alkali feldspar and foid.

Geochemical characteristics of rocks   The average chemical compositions (in %) of subalkaline basaltic rocks are as follows: for picritic basalt, SiO_2 43.57 ± 1.14, AlK 1.95 ± 0.47, K_2O 0.51 ± 0.37, Na_2O 1.44 ± 0.57 and MgO up to 13.28−19.05 (in Mg − rich varieties); for basalt, SiO_2 49.45 ± 1.61, AlK 3.09 ± 0.49, K_2O 0.52 ± 0.45 and Na_2O 2.57 ± 0.53. The average chemical compositions (in %) of alkaline basaltic rocks are as follows: for basanite / tephrite, SiO_2 44.34 ± 1.67, AlK 5.40 ± 0.97, Na_2O 3.47 ± 0.93 and K_2O 1.93 ± 1.02; for phonolitic basanite / tephrite, SiO_2 49.03 ± 1.50, AlK 8.83 ± 1.7, Na_2O 4.62 ± 1.58 and K_2O 4.21 ± 1.93; for tephritic phonolite, SiO_2 53.64 ± 2.00, AlK 11.27 ± 2.38, Na_2O 5.88 ± 2.38 and K_2O 5.69 ± 2.29; for foidites SiO_2 40.05 ± 2.98. Subalkaline basalt contains (in ppm) 3.6−17.0 Rb, 520−750 Sr, 70−235 Ba, and 9.4−47 Cr, 8.2 − 36 Ni and 6−14 Co; tholeiite has $\sum$ REE = 35.56−94.57 and La / Yb = 0.82−2.59; calc − alkaline basalt has $\sum$ REE = 31−70 and La / Yb = 3−8; subalkaline basalt has $(^{87}Sr/^{86}Sr)_i$ = 0.7028−0.7034. Alkali basalt contains 297−359 Cr and 142−263 Ni; sodic varieties have $\sum$ REE = 185 − 322 ppm and La / Yb = 9.8 − 34.9; potassic varieties have $\sum$ REE = 280−643, La / Yb = 47−74 and $(^{87}Sr/^{86}Sr)_i$ = 0.7033−0.7045.

TEMPORAL−SPATIAL DISTRIBUTION OF IGNEOUS ROCK ASSOCIATION

Igneous rocks of different ages in eastern China are widespread, covering an area of 1.82 million km^2 (accounting for 19% of China's land area), of which volcanic rocks make up 9.3% and plutonic rocks 9.7%. Among the volcanic rocks, basic rocks account for 51%, while intermediate ad acid rocks 49%. Among the plutonic rocks, intermediate and acid rocks account for 97.8%, basic and ultrabasic ones 2.1% and alkaline ones 0.1%. Three tectonic settings may be broadly distinguished, i.e.: Precambrian active belts, Paleozoic to early Mesozoic composite orogenic belts and Meso – Cenozoic continental activation belts. The temporal – spatial distribution characteristics of igneous rock associations therein are here described separately as follows.

Temporal–Spatial Distribution of Precambrian Igneous Rock Associations
Precambrian igneous rocks of China are mainly exposed in the basement uplift areas of the North China, Tarim and Yangtze blocks. Magmatism generally occurred in five stages: Archean, Early Proterozoic, Middle Proterozoic, Late Proterozoic Qingbaikou'an and Late Proterozoic Changchengian. Except migmatite in the paleocontinental nucleus, igneous rocks are mostly distributed in paleocontinental block (called block for short hereinafter) breakup zones, block accretion margins and zones of interaction of different blocks. (1) In block extension and breakup zones, there mainly appear metasedimentary rocks (schist and marble) and metamorphosed basaltic lavas of the tholeiitic suite and pyroclastic rocks (amphibolite and granulitite); (2) on block accretion margins and in zones of interaction and compression of different blocks, there mainly occur metamorphosed basaltic – andesitic – rhyolitic rock associations (amphibolite, amphibole – plagioclass gneiss and granulitic) of the calc – alkaline suite and associations of tonalitic, granodioritic and plagiogranitic plutonic rocks (grey gneisses). (1) Archean igneous rocks are mainly concentrated in the high – grade metamorphic series of the northern and eastern parts of the North China blocks and northern part of the Tarim block, and the meta – igneous rock associations reflect the features of extension, compression and again extension. (2) Early Proterozoic volcanic rocks are mainly distributed on the northwestern margin of the Tarim block, northern and southern margins of the North China block and the southern margin of the Qaidam block as well as on the western margin of the Yangtze block. They consist dominantly of metamorphosed subalkaline basaltic, andesitic, dactic and rhyolitic calc – alkaline rocks reflecting the features of epicontinental accretion. On the local margins of the blocks, there appear alkaline volcanic rocks related to extension, accompanied by the corresponding tonalite, granodiorite and plagiogranite association and granodiorite, quartz monzonite and granite association. (3) Middle Proterozoic volcanic rocks are dominated by epicontinental accretionary calc – alkaline rocks on the northern margin of the Tarim and North China blocks and there appear bimodal volcanic rock associations related to extension on the southern margin of the North China block; in the Central Qilian and North Qinling mountains, epicontinental accretionary calc – alkaline volcanic rock associations predominate, and in the South Qinling Moutnains there occurred bimodal volcanic rocks of the extensional environment. Plutonic rocks of this age are represented by the tonalite, granodiorite and plagioclase granite association and the granodiorite – monzogranite – granite association, in addition to peraluminous granite and alkali syenite appearing in local areas, e.g. Fanjingshan of Guizhou and Wangcang of Sichuan, and basic and ultrabasic associations appearing locally on the northern margin of the Tarim block and North China block. (4) Late Proterozoic volcanic rocks are represented by calc – alkaline volcanic rocks in most areas, and bimodal volcanic rocks occur only in some areas, e.g. Wuyi – Yunkai. In most cases plutonic rocks are marked by the granodiorite – monzogranite – granite association, the peraluminous association being only found in some areas such as Xuefeng – Jiuling and the alkali feldspar granite and alkali granite association appearing in some areas, e.g. western Sichuan and eastern Yunnan.

Temporal—Spatial Distribution of Paleozoic Igneous Rock Associations
China's Paleozoic igneous rocks are generally concentrated in composite orogenic belts, e.g. the Tianshan — Hinggan and Qilian — Qinling orogenic blets. The plutonic rocks are more widespread, e.g. in the Nanhua tectonic belt. Early Paleozoic magmatism began in the Early Cambrian and ended in the Late Silurian, but volcanism was the strongest in the Ordovician. Late Paleozoic magmatism occurred more extensively.

The geological settings of Paleozoic igneous rocks associations may be summarized as follows: (1) oceanic ridges or relict fragments of oceanic crust in assembly zones of different blocks; (2) subducted volcanic or magmatic arcs; (3) collisional volcanic or magmatic arcs; (4) intra — arc or interarc grabens; (5) back — arc extensional basins; (6) large — scale strike — slip or transform fault zones; (7) mobile belts on margins of median massifs; (8) rifts or aulacogens on block margins; (9) rifts or hot — spot areas in the block interior.

The igneous rock associations representing different tectonic settings show certain distribution patterns in space and time and tend to appear in association with each other. Belts of basic volcanic rocks and intrusive rocks as a part of tectonic emplacement ophiolites occurring in assembly zones of different blocks include the Early Paleozoic Fuyun — Beitashan belt, Tangbale — Ondor Sum — Xar Moron River belt, North Qilian belt and South Qilian belt, the Late Paleozoic Kelamaili — Moqin Ul belt, Dalabud — Hegen Mts. belt, Bayan Gol — Solon Mts. belt, West Kunlun northern belt and East Kunlun southern belt, and the Paleozoic Qinling belt.

The associations of volcanic rocks and intrusive rocks occurring as subducted arcs are mainly developed above the transitional crust. The immature island arc developed at the setting of relatively thin continental crust consists mainly of the subalkaline basalt and basaltic andesite association; the continental — margin arc formed at the setting of relatively thick continental crust consists dominantly of calc — alkaline andesitic rocks; in addition there usually occur considerable amounts of dacitic and rhyolitic rocks. In the process of island arc development there occurred locally transient extension, forming a bimodal volcanic rock association representing an intra — arc graben. In the back — arc basin there may appear the olivine tholeiite association associated with normal sedimentary rocks. The basaltic trachyandesite — trachyandesite association, trachydacite (or high — K dacite) and high — silica rhyolite association may appear at the late — stage mature or postmature island arcs, especially those related to large strike — slip or transform faults. Representative volcanic or magmatic arc belts include the Early Paleozoic Central Tianshan northern margin — Duobaoshan belt (O — S), Boin Sum — Shifeng belt (O_2) and Qilian Mts. — North Qinling belt (S), and the Late Paleozoic Central Tianshan southern margin belt (D — C), Central Tianshan northern margin belt (C_1–C_2) and Lancang River belt (P_1–P_2).

The associations of volcanic and intrusive rocks occurring as collisional arcs display more complex tectonic relationships in addition to the essential attributes of the subducted arcs, and the igneous rock associations also have diverse features, which might be related to continental crustal thickening caused by collision and orogeny.

The volcanic and intrusive rocks formed in the collision stage consist essentially of intermediate − acid rocks of the calc − alkaline suite, including the tonalite, monzogranite and plagiogranite association and the biotite granite, two − mica granite and muscovite granite association, the latter being alumina − oversaturated. Late Paleozoic collisional igneous rocks are commonly distributed between the Siberian and Tarim plates, from north of Mt. Hantengri and Muzat Daban on the southern margin of the Central Tianshan through Xingxingxia and Erenhot to Xilinhot. Their age is mainly Middle − Late Devonian. Early Paleozoic alumina − oversaturated granite is mainly distributed in the Yunkai, Wugong and Wuyi mountains. Late − stage collisional and postcollisional igneous rocks are marked by an association of gabbro, diorite, quartz diorite, granodiorite, alkali feldspar granite and alkali granite and their volcanic equivalents as well as basaltic trachyandesite and trachyandesite and their intrusive equivalents, occurring in West Junggar and Shitoluogai (C_3) and Erenhot − Xilinhot of Inner Mongolia (Nei Mongol). Rift basalt related to continental − margin extension is represented by Late Paleozoic E'mei flood basalt (P_2) and lamproite in Guizhou and Hunan. Alkaline rocks, alkali ultrabasic rocks and carbonatite related to intracontinental extension or hot spots (mantle plumes) mainly include alkali syenite and carbonatite in Nanjiang, Sichuan, and kimberlite ($\in$ −O) in Mengying of Shandong, Fu Xian of Liaoning and Zhenyuan of Guizhou. On active margins of median massifs in composite orogenic belts there often occur some intrusive rock belts, dominated by hypabyssal intrusive rocks of granodioritic, quartz monzonitic and granitic compositions, as exemplified by those on peripheries of the Qaidam massif and on the western margin of the Jiamusi massif.

Temporal−Spatial Distribution of Meso−Cenozoic Igneous Rock Associations
Meso−Cenozoic igneous rocks are mainly distributed in the peri − Pacific region of eastern China and less commonly in southwestern China, especially in the Neo − Tethys tectonic domain.

In terms of the geological environments of Meso − Cenozoic igneous rocks, subducting and collisional orogens are represented in Taiwan and the Zangdian (Tibet − Yunnan) region, especially the Sanjiang area where secondary tectonic units are completely developed and the types of igneous rock association are complete too. The peri − Pacific region of eastern China is marked by continental activation belts, where the main geological environments of Meso − Cenozoic igneous rocks include: (1) active margins of the basement uplift area; (2) superposed down − faulted basins of the basement uplift area; (3) active margins on the uplifted side of a large − scale strike − slip fault; (4) inherited basins on the depressed side of a large − scale strike − slip fault; (5) epicontinental extension grabens; (6) intracontinental rift and aulacogens.

The associations of volcanic and intrusive rocks representing different geological environments show certain patterns of distribution in time and space and can occur in company. In terms of time, magmatism mainly occurred in the Sanjiang area during the Late Permian to Triassic, in Tibet in the Meso − Cenozoic and in eastern Taiwan in the Cenozoic. In the peri − Pacific region in eastern China magmatism occurred in three stages: (1) the post − collision (P_2−T), when nearly E − W − trending lag − type magmatic activity belts were formed; (2) the stage of transformation of the regional stress field (J − K), when there occurred NE − and NNE − trending tectono — magmatic belts in which volcanic and intrusive rocks related to large — scale strike — slip faults were developed; (3) extension stage (E–Q), when there were developed NNE — trending, rift— and aulacogen — type volcanic belts.

On the basis of the geological settings of igneous rocks and their associations, the peri — Pacific region in eastern China may be divided into 3 rock provinces, 10 rock belts and 21 rock districts.

Northeast China Mesozoic igneous rock province Volcanic and intrusive rocks of this province are mainly distributed in Paleozoic fold belts on the southern margin of the East Siberian platform and a part of them occur on acitve margins of median massifs between fold belts. They may fall into four rock belts and five rock districts. The rocks are characterized by the development of associations of high — K dacitic — rhyolitic rocks and their intrusive equivalents with varying amounts of basaltic, andesitic and trachuandesitic rocks and their intrusive equivalents (Table 1).

Table 1. Associations of Mesozoic igneous rocks in Northeast China

Rock belt	Rock district	Volcanic rock association	Intrusive rock association	Geological setting
Hinggan	Northern Hinggan	J_3–K_1, high–K dacitic–rhyolitic with minor basaltic	J_3–K_2, granodiorite–granite	Paleozoic fold belt
	Southern Hinggan	J_2–K_1, high–K dacitic–rhyolitic with some of basaltic andesitic and trachyandesitic	J_2–K, quartz diorite–granodiorite and monzogranite	Late Paleozoic fold belt with ophiolite
Xilinhot	Wanda	T_2–J_1, basaltic–basaltic andesitic	T_1, granodiorite–tonolitic association with ultramafic pyrolite and diabase	Exotic nappe outlier of oceanic crust
Eastern Jilin–Eastern Heilong-jiang	Eastern Jilin–Eastern Heilong-jiang	J_3–K_1, andesitic–trachyandesitic–trachydacitic–rhyolitic (bimodal association appears during the late stage)	T_3–K, quartz diorite–granodiorite–monzogranite–granite	Early Paleozoic fold belt
Southern Songliao	Southern Songliao	J_3–K_1, trachybasaltic–trachyandesitic–trachydacitic–rhyalilic	T_3–J_1, monzonite–quartz monzonite–monzogranite	Late Paleozoic fold belt

Volcanic and Intrusive rocks of the Northeast China Mesozoic rock province are of inhomogeneous distribution. In the Da Hinggan Mountains area in the west part, volcanic rocks predominate, limited intrusive rocks being exposed; the rocks are less eroded. In the east part, intrusive rocks are dominant and volcanic rocks are subordinate. The distribution of the Da Hinggan Mountains area is principally controlled by the NNE– and NE — trending fault belts. Their northern and southern sectors both consist dominantly of trachydacite (or high–K dacite) and rhyolite, but there appear more basaltic, andesitic and trachyandesitic rocks in the southern sectors and there are less intermediate and basic volcanic rocks in the northern sectors than in the southern sector. In the east, intrusive rocks are generally distributed in a NW direction along the boundaries of interactions of different secondary blocks; less commonly intrusive and volcanic rocks are controlled by the nearly E — W — trending faults along the northern boundary of the North China platform, and besides, intrusive and volcanic rocks also occur along the NNE — trending faults.

North China Mesozoic igneous rock province   Mesozoic volcanic and intrusive rocks of the North China platform are mainly distributed on its northern margin, near the Tanlu (Tancheng — Lujiang) fault and in the Taihang and Lüliang fault zones parallel to the Tanlu fault. Three rock belts and eight rock districts may be distinguished (Table 2). As shown in Table 2, as the basement of the North China platform is composed of special crustal and mantle materials, its representative volcanic and intrusive rock associations have much similarity, i.e. intermediate volcanic and intrusive rocks dominant, rich in alkalis and polassium to different degrees, forming associations composed mainly of trachybasaltic, basalt trachyandesitic, trachyandesitic, trachytic and even phonolitic rocks or of a part of the aforesaid rocks. Such associations were formed in different stages from the Triassic to Cretaceous and in different areas, reflecting the similarity between the material composition of the lower crust and that of the upper mantle of the North China block and geochemical inheritance of magma generated by partial melting. Such intermediate — basic, high — alkali and high — potassium magmatism was related to $T-J_2$ and J_3-K_1/K_2 large strike — slip fault activities. On the northern margin of the North China block, the distribution of $T-J_2$ igneous rocks is controlled of faults of nearly E–W trends, with basic and intermediate volcanic rocks mainly occurring in down — faulted areas and intrusive rocks occurring in uplifted areas of the Precambrian metamorphic basement. J_3-K_1- igneous rocks are mainly related to NNE — trending large — scale strike — slip faults and their derivative faults, mainly distributed in the Tanlu fault zone and in western Shandong, the southern Taihang Mountains and Lüliang Mountains. Volcanic and intrusive rocks are generally distributed in a NNE or nearly N–S direction, and the starting time of magmatic activities became late gradually from east to west, while their ending time was close. Large numbers of inclusions representing the rocks of the upper part of the upper mantle and the lower part of the lower crust have been found in this type of intermediate — basic to intermediate rocks. Petrology and trace element and isotope geochemistry have proved that the magma of this type of igneous rocks was derived from the upper part of the upper mantle and the lower part of the lower crust. The differences in material composition of the crust and mantle, degree of partial fusion and degree of differentiation further complicate the associations and geochemical characteristics of igneous rocks. As the ascent of basic and intermediate magmas and concomitant upward migration of the deep — seated heat flow — peak region also resulted in anatexis of metamorphic rocks of teh middle crust, there originated significant amounts of trachydacitic — rhyolitic volcanic rocks and granodioritic — quartz monzonitic — monzonitic — granitic intrusive rocks; especially near sea areas, the background field of the regional heat flows is relatively high and the ascending part of the heat flow peak region is relatively shallow, so granitic bodies are better developed in the eastern part of the North China platform than in the western part.

Table 2. Mesozoic igneous rock associations in North China

Rock belt	Rock district	Volcanic rock association	Intrusive rock association	Geological setting
Northern margin of North China	Western sector	J_3, trachyandesitic–trachydacitic–rhyolitic	P_2–T, monzonite–monzosyenite with minor alkali syenite	Precambrian basement uplifted area (with Archean basement)
	Middle sector (Yanliao)	J_1, basaltic–andesitic; J_2, andesitic–trachyandesitic; J_3–K_1, trachyandesitic–trachydacitic–dacitic	J_3–K_1, diorite–granodiorite–monzogranite–granite	Proterozoic metamorphosed basement, related to a large scale–strike slip faults
	Eastern sector	T, trachytic–phonolitic	T–J_2, syenite–alkali syenite–diorite–monzonite	Middle–Late Proterozoic metamorphosed basement, related to faulting
Shandong	Wester Shandong	J_3–K_1, basaltic trachyandesitic–trachyandesitic–trachytic, andesitic–trachyandesitic	J_3–K_1, gabbro–gabbrodiorite–monzonite, monzonite–syenite, monzogranite–granite	Archean and Proterozoic metamorphosed basement, related to fault system
	Tanlu	J_3–K_1, basaltic trachyandesitic–trachuandesitic–trachytic, trachyandesitic–trachytic–phonolitic	J_3–K_1, monzogabbro–monzonite, quartz monzonite–monzogranite / quartz syenite	Precambrian basement and large–scale strike–slip faults
	East Shandong	J_3–K_1, andesitic–dacitic–rhyolitic, trachyandesitic–trachydacitic–rhyolitic	J_3–K_1, diorite–granodiorite, quartz monzonite–monzogranite–granite	Precambrian basement in the eastern side of the Tanlu faults
Shanxi	Wutai–Taihang	J–K, andesitic–trachyandesitic and subvolcanic (Wutai)	J_1–K_1, diorite–monzonite and monzonite–syenite–alkali syenite	On the uplift side of a large–scale strike–slip fault, with the Precambrian metamorphic basement
	Lüliang	J_3–K_1, alkali volcanic rock (trachytic, and phonolitic	J_2–K_1, diorite–monzonite and syenite–alkali syenite	Precambrian metamorphic basement, related to faults

South China Mesozoic igneous rock province In contrast with North China, Mesozoic volcanic rocks in South China are represented by associations of trachydacitic – rhyolitic or high–K dacitic – rhyolitic rocks, sometimes with small amounts of trachyandesitic or quartz tranchyandesitic rocks; intrusive rocks are represented by the I—type granitic — quartz monzonitic — granitic association, sometimes with small amount of gabboic — dioritic and monzonitic — dioritic rocks. In areas represented by the Nanling Mountains there occur intrusive rock associations represented by the S—type biotite granitic — two — mica granitic — muscovite granitic rock association with small amount of corresponding peraluminous dacitic and rhyolitic rocks, characterized by the presence of Al — rich and volatile — rich minerals (garnet, cordierite, topaz etc.). The South China rock province may be divided into three rock blets and nine rock districts (Table 3). There are many blocks in South China and various blocks differ somewhat in structure and composition of the crust and mantle as well as geochemical features; so the types and characteristics of volcanic and intrusive rocks formed arc quite different. For example, I— and A—type igneous rock associations predominate in eastern Zhejiang and the coastal area of eastern Fujian, I—type igneous rock associations is dominant in the middle — lower Yangtze River valley and the northeastern Jiangxi — northwestern Zhejiang zone, peraluminous S—type associations are dominant in Hunan, Guangxi and Guangdong (western Guangdong), and I— and S—type or transitional type associations are represented in southern Anhui and northern Jiangsu.

Cenozoic volcanic rock province   Cenozoic volcanic series mostly occurred along Mesozoic NNE — trending great faults, which were subjected to further extension to form aulacogens or rift belts. These volcanic belts mostly cross the Northeast China, North China and South China Mesozoic igneous rock provinces. The volcanic rock associations may fall into three types, i.e. (1) volcanic rock associations mainly of tholeiite, including olivine tholeiite, tholeiite and quartz tholeiite, sometimes with small amounts of alkali basalt, whose symbol is TH$-\beta$; (2) volcanic rock associations mainly of alkali basalt, which may be divided into the potassic ones and sodic ones: the potassic ones consist of alkali picritic basalt, tephrite (or basanite), tephritic (or basanitic) phonolite and leucite phonolite and the sodic ones also alkali picrobasalt and tephrite (or basanite) in addition to phonolitic basanite and foidite (nephelinic or leucite — nephlinic), with A$-\beta$ as their symbol. (3) associations of alkali basaltic, especially sodic basaltic, rocks and tholeiitic rocks, with Aβ—THβ as the symbol. The third type occurs in most cases. Sometimes there appearred some alkali trachytic, basanitic and alkali — rhyolitic nimbrites in the late stage of differentiation of sodic basaltic magma, marked by the symbol A$-\lambda$. Cenozoic volcanic rocks in China may be divided into three rock provinces, with Northeast China and North China merging into one rock province and the coastal region of southeastern China and Taiwan being separate rock provinces. Each rock province may be further divided into 9 rock belts and 39 rock districts on the basis of their different stages of formation and modes and areas of occurrence of basalt. Mantle peridotite occurs in large amounts in Cenozoic basalt, particularly in sodic picrite basalt and basanite. It includes I—type (i.e. lherzolite) inclusions (or called green — type inclusions) and II—type (i.e. wehrlite) inclusions (or called black — type inclusions). The I type consists dominantly of spinel lherzolite with small amount of garnet lherzolite, mica lherzolite, hornblende lherzolite and plagioclase lherzolite. The II—type includes not only inclusions representing pyrolite but also pyroxenite and gneiss inclusions derived from the lower crust. They respectively represent inclusions of country rocks captured in the source regions of partial melting of magmas and their passageways and have important significance for the further study of the compositions of the upper mantle and lower crust in the study region. Eastern China is a continent assembled by multiple blocks, different blocks of the

Table 3. Associations of Mesozoic igneous rocks in South China

Rock belt	Rock district	Volcanic rock association	Intrusive rock association	Geological setting
Tongbai–Dabie	Tongbai areas	J_3–K_1, trachyandesitic–trachytic; andesitic–dacitic– rhyolitic	J_{2-3}, monzonite–quartz monzonite– granite	Pre–Sinian metamorphic basement, on the northeastern part of the Wudang uplift (in the western side of Tanlu faults)
	Dabie area		K_1, granodiorite–quartz monzonite–monzogranite	
	South Shandong	J_3–K_1,trachybasaltic–basaltic trachyandesitic–trachyandesitic (bimodal volcanic rocks occur only during late stage)	Granodiorite–monzogranite– moyite–K–feldspar granite	Pre–Sinian metamorphic basement (on the eastern part of Tanlu fault zone)
Yangtze blocks	Lower Yangtze	J_3–K_1, trachybasaltic–basaltic trachyandesitic–trachyandesitic–trachytic–phonolitic, andesine basaltic–basaltic trachyandesitic–trachyandesitic–trachydacitic	J_3–K_1, quartz diorite–granodiorite–quartz monzonite–monzogranite, syenite–quartz syenite–granite	On the Paleozoic down–faulted basins and local uplifted areas of Precambrian basement
	Jiangnan block	J_3–K_1, trachydacitic rhyolitic; dacitic–rhyolitic, minor trachyandesitic or andesitic	J_2–K_1, quartz diorite–granodiorite; quartz monzonite–monzogranite	Proterozoic metamorphic basement
Nanhua rock belt	Gannan–Minxi (South Jianaxi–West Fujian)	J_1–J_2, peraluminous andesitic–trachyandesitic–trachydacitic–rhyolitic	T–J_1, quartz diorite–granodiorite–quartz monzonite–monzogranite; peraluminous binary granite–muscovite granite	Lower Paleozoic fold belts
	Zhedong Mindong (the Eastern Zhejiang–Fujian	J_3–K_1, high–K dacitic–rhyolitic, with minor trachyandesitic	J_3–K_1, granodiorite–quartz monzonite–monzogranite–granite-ulkali feldspar granite	Proterozoic and Early Paleozoic folded basement
	Yuexi (West Guangdong)–Hainan	T_1–J_1, bimodal association of basaltic–rhyoritic; J_3–K_1, high–K dacitic–rhyolitic	T_1–J_1, gabbro–granite; J_3–K_1, granodiorite–quartz monzonite–granite	Late Paleozoic folded basement

continent have different compositions of the crust and mantle and in particular the degrees of depletion of incompalible elements in the mantle are different, so there are great differences in trace element and isotope geochemical feature though the same rock associations and names are designated.

RELATIONS OF MAGMA TYPES WITH THE STRUCTURES AND COM-POSITIONS OF THE CRUST AND MANTLE

The types of primary magma represented by Mesozoic igneous rock associations in eastern China have intimate relations with the depths and compositions of their crust and mantle sources, which are described as follows.

Acid Igneous Rock Association Derived from the Boundary between the Upper Crust and Middle Crust
The integrated inverse magnetic, gravity and seismic profiles indicate that low − velocity layer of seismic waves corresponding in depth with the region of peraluminous acid igneous rocks is located in the zone of interaction or tectonic detachment between the top of the middle crust and the lower part of the upper crust. The igneous rocks are represented by the association of peraluminous rhyolitic − dacitic and silica − rich rhyolitic rocks and the association of biotite granitic, two − mica granitic and muscovite granitic rocks. Hydrated ferromagnesian minerals of intrusive rocks include annite, siderophyllite and muscovite with ilmenite and Al−rich and volatile − rich accessory minerals. Generally rock inclusions are absent. The SiO_2 content$= 70\% − 76\%$; $K_2O/(Na_2O+K_2O)= 0.48 − 0.60$; $Al_2O_3/(CaO+Na_2O+K_2O)* = 1.03 − 1.26$; $Mg/(Mg+Fe+Al+Ti+M)* = 0.10 − 0.25$; $Rb/Sr>20$; $\delta Eu<0.20$. The low − grade metamorphic rocks in magma source regions are mostly Middle Proterozoic argillaceous − arenaceous, fine clastic metasedimentary rocks. In terms of isotope geochemical features, the data points are located far from the mantle line (with a high μ value) on the $^{206}Pb/^{204}Pb-^{207}Pb/^{204}Pb$ diagram and on the left side of the mantle line and near it on the $^{206}Pb/^{204}Pb-^{208}Pb/^{204}Pb$ diagram; the ^{204}Pb content is $<1.36\%$ and $^{207}Pb/^{206}Pb$ and $^{208}Pb/^{206}Pb$ are lower than 0.85 and 2.10 respectively; $(^{87}Sr/^{86}Sr)_i>0.712$, $\varepsilon_{Nd}(t)<10$ and $\delta^{18}O>+10‰$. This type of magma was produced by anatexis and selective fusion of materials at the base of the upper crust and in the upper part of the middle crust under the water− and volatile − rich conditions.

Intermediate−Acid and Acid Igneous Rock Association Derived from the Middle Part of the Middle Crust
The integrated inverse geophysical profiles indicate that the low − velocity layer corresponding in depth to the region of intermediate − acid and acid igneous rock with higher Al contents is located in the zone of interaction or tectonic detachment between the upper and lower velocity layers of the middle crust. Igneous rock associations are dominated by the association of high − alkali dacitic and rhyolitic volcanic rocks and the association of granodioritic and biotite granitic intrusive rocks. The ferromagnesian minerals of the intrusive rocks are mainly hornblende and biotite with magnetite − sphene type accessory minerals. The rocks contain a large amount of dioritic inclusions. $SiO_2= 61-73\%$; $K/(K+Na)<0.42$; $Al/CNK <0.98$; $M/MF>0.30$; Rb/Sr (for intermediate − acid rocks) <0.8 and Rb/Sr (for acid rocks) <1.2; $\varepsilon Eu= 0.8 − 1.1$. The source rocks are mostly marked by Late

Archean or Early Proterozoic metavolcanic rocks, belonging to the medium − to medium − high − grade metamorphic terrains. Data points lie immediately above the mantle line on the $^{206}Pb/^{204}Pb-^{207}Pb/^{204}Pb$ diagram but on the left of the Lower upper crust line on the $^{206}Pb/^{204}Pb-^{208}Pb/^{204}Pb$ diagram. The ^{204}Pb content ranges between 1.36 and 1.42‰; $^{207}Pb/^{206}Pb$ mostly ranges between 0.86 and 0.92; $^{208}Pb/^{206}Pb$ is 2.12 − 2.29; $(^{87}Sr/^{86}Sr)_i$ < 0.712; $\varepsilon_{Nd}(t)$ < − 8; $\delta^{18}O$ = +8 to +11‰.

Intermediate−Acid and Acid Igneous Rock Associations Derived from the Boundary between the Middle and Lower Crust

Geophysical profiles indicate that the deep low − velocity layer corresponding to the region of this type of high − K igneous rocks is situated between the lower part of the middle crust and the upper part of the lower crust, corresponding to the zone of interaction or tectonic detachment between two velocity layers. The igneous associations consist dominantly of high − K intermediate − acid and acid igneous rocks with small amounts of intermediate or even basic igneous rocks. The common associations are the association of intrusive rocks mainly of granodioritic, quartz monzonitic, monzogranitic and alkali feldspar granitic (± alkali granitic) compositions and the corresponding volcanic rock association mainly of quartz trachyandesitic, trachydacitic and high − K and high − silica rhyolitic compositions; besides, minor amounts of gabbroic, gabbro − monzonitic and monzonitic intrusive rocks or basaltic and basalt − trachyandesitic rocks are present and their area is generally less than 10% of the total area of intrusive complexes or volcanic complexes. Acid intrusive rocks are characterized by the appearance of magnesian biotite; intermediate intrusive rocks may contain two kinds of pyroxene and accessory minerals are mainly magnetite and sphene. The end − member rocks of the intermediate and basic rocks contain varying amounts of dark − colored inclusions; the acid rocks commonly contain no inclusions. Alkali granite usually contains miarolitic cavities, and sometimes there appear materials differentiated from alkali granite which contain alkali ferromagnesian materials. The SiO_2 content ranges from 64 − 77% but may attain 50 − 56% in intermediate − basic rocks; K/(K+Na)<0.48; Al/CNK<1; M/MF = 0.2−0.25; Rb/Sr = 0.5 − 2; δEu = 0.2 − 0.9. Data points fall close to the mantle line on the $^{206}Pb/^{204}Pb-^{207}Pb/^{204}Pb$ diagram and far from the mantle line on the $^{206}Pb/^{204}Pb-^{208}Pb/^{204}Pb$ diagram; ^{204}Pb > 1.4%, $^{207}Pb/^{206}Pb$ > 0.90 and $^{208}Pb/^{206}Pb$ > 2.18; $(^{87}Sr/^{86}Sr)_i$ for acid rock < 0.710 and that for the basic end member < 0.706; $\varepsilon_{Nd}(t)$ < −10; $\delta^{18}O$ < +10‰.

Intermediate−Basic and Intermediate Igneous Rock Associations Derived from the Boundary between the Crust and Mantle

The deep low − velocity layer corresponding to the region of this type of high−K intermediate − basic and intermediate igneous rocks is located in the lower part of the lower crust and at the top of the upper mantle, corresponding to the zone of interaction or tectonic detachment between the crust and mantle. In comparison with the seismic wave − velocity background field between the lower crust and upper mantle in the surroundings there appear marked local anomalies. The intrusive rock associations are mainly of gabbro − monzonitic and monzonitic compositions. The basic end member is gabbroic or hornblende gabbroic; the alkalic differentiated end member is syenitic or alkali syenitic; the acid differentiated end member is quartz syenitic or monzogranitic. The volcanic rock associations are dominated by basalt − trachyandesitic and trachyandesitic ones, the basic end member being trachybasaltic or andesine basaltic, the alkalic end member being trachytic and phonolitic and the acid end member being quartz trachyandesitic and trachydacitic. The ferromagnesian minerals are characterized by water– and volatile – rich ones. There usually appear hornblende or biotite and two kinds of pyroxene. Two kinds of feldspar are common and their content ratio varies with lithology. The accessory minerals are magnetite and apatite. Apatite is rich in volatile fluorine. There occur varying amounts of inclusions from the upper part of the upper mantle and lower crust and xenoliths from different depths in the igneous rocks of the basic end member. The Na_2O and K_2O contents are relatively high. According to the ratio of the relative contents of both, the potassic rocks with $K_2O > (Na_2O-1.5)$ and sodic rocks with $K_2O < (Na_2O-1.5)$, but according to the proportions of K_2O and SiO_2 the rocks mostly belong to high–K ones. $M/MF > 0.35$ and $Rb/Sr < 1.0$. Data points plot immediately below the mantle line on the $^{206}Pb/^{204}Pb-^{207}Pb/^{204}Pb$ and are deviated from the mantle line on $^{206}Pb/^{204}Pb-^{208}Pb/^{204}Pb$; $^{204}Pb > 1.40\%$, $^{207}Pb/^{206}Pb < 0.90$, $^{208}Pb/^{206}Pb > 2.18$, $^{87}Sr/^{86}Sr < 0.708$ and that for the basic end member < 0.706; $\varepsilon_{Nd}(t) < -10$ and $\delta^{18}O < +10‰$.

Cenozoic rift – type basalts in eastern China may be classified into five types on the basis of their rock associations, deep – seated inclusions and Sr–Nd–Pb isotope geochemical features. They are described separated as follows.

Isotope geochemistry of the depleted mantle – type basalt association   China's Cenozoic basalts derived from the end member of the depleted mantle are mainly represented by the sodic alkali – basaltic and tholeiitic volcanic rock association. Alkalic basaltic rocks include alkali olivine basalt and basanite, sometimes with small amount of olivine nephelinite; while tholeiite comprises olivine tholeiite, tholeiite and quartz tholeiite. The former commonly contain varying amounts of ultramafic inclusions, composed mainly of spinel lherzolite with some spinel – olivine websterite. Basalt associations formed by the upper mantle and asthenospheric mantle depleted to different degrees are mainly distributed in South China, e.g. the South China Sea basin, Leiqiong area, Guangdong, Zhejiang and some areas of central Fujian as well as some areas on peripheries of the Bohai Sea. Generally speaking, the basalt association representing the depleted mantle end member of China has the following isotope geochemical features: $^{87}Sr/^{86}Sr = 0.7029-0.7032$, $^{143}Nd/^{144}Nd = 0.51310-0.51385$, $^{206}Pb/^{204}Pb = 18.11-18.60$, $^{207}Pb/^{204}Pb = 15.47-15.65$ and $^{208}Pb/^{204}Pb = 38.15-38.78$.

As the Sr–Nd–Pb isotope system of the lithospheric mantle is inhomogeneous in the same area and its nature may change with time, the mantle may display the features of enrichment locally in a particular time even if it generally has the features of depletion. Things are not absolute and remain unchanged for ever. What is important is to judge which plays the dominant role, depletion or enrichment.

Isotope geochemistry of the enriched mantle–type basalt association   China's Cenozoic basalts derived from the enriched mantle end member are represented by K–rich alkali basalt in Wudalianchi, Keluo and Erke Mountain and secondarily by tholeiite of the Beijing depression. The basalt association of Wudalianchi is composed of alkali picrite, phonolitic basanite, basanite phonolite, phonolite and leucitite, which are characterized by the presence of varying amounts of leucite. The representative is the Keluo leucite basalt. Its distinctive features are the enrichment of the Sr, Nd and Pb systems and particularly the very high content of incompatible elements and radioactive Sr. The Eogene tholeiite association, including olivine tholeiite, tholeiite and quartz tholeiite, represents the moderately enriched mantle source. It has a relatively low content of incompatible elements and is moderately enriched in radioactive Sr isotopes and slightly enriched in Nd isotopes.

The enriched mantle end member represented by the Wudalianchi alkali basalt association has the following isotopic features: $^{87}Sr / ^{86}Sr = 0.70512-0.70602$ with a mean of 0.70539, $^{143}Nd / ^{144}Nd = 0.512381-0.52433$, $^{206}Pb / ^{204}Pb = 16.712-17.310$, $^{207}Pb / ^{204}Pb = 15.333-15.490$ and $^{208}Pb / ^{204}Pb = 36.259-37.338$. The isotopic values of the moderately enriched mantle source areas represented by Eogene tholeiite are as follows: $^{87}Sr / ^{86}Sr = 0.70423 - 0.70528$ and $^{143}Nd / ^{144}Nd = 0.51267 - 0.51296$. The most depleted quartz tholeiite in the Beijing depression has $^{206}Pb / ^{204}Pb = 16.858$, $^{207}Pb / ^{204}Pb = 15.294$ and $^{208}Pb / ^{204}Pb = 36.829$. The isotopic composition of basalts has many constraints though it essentially shows the features of the isotopic composition of the mantle of the source area. So it is necessary analyze the Sr—Nd—Pb isotope system as a whole, and the degree of enrichment of the mantle reflected represents a basic trend. In fact the three — dimensional inhomogeneous microscopic change of the composition of the upper mantle is rather complex. The metasomatism of the mantle plays an important role.

Isotope geochemistry of the quasi—primitive mantle—type basalt association   Basalts with the quasi — primitive mantle — type isotopic composition are mainly distributed in most parts of North China and a part of eastern China, excluding some local concentration areas and local depletion areas. The volcanic rock associations consist mainly of alkali basaltic and tholeiitic ones. The alkali basalt association is dominantly composed of alkali olivine basalt, basanite and phonolitic basanite, sometimes with small amount of foiddite. The tholeiitic association includes olivine tholeiite, tholeiite and quartz tholeiite.

The isotopic values of this type of basalt are as follows: $^{87}Sr / ^{86}Sr = 0.70380-0.70497$, $^{143}Nd / ^{144}Nd = 0.512627-0.5128$, $^{206}Pb / ^{204}Pb = 17.175-17.951$, $^{207}Pb / ^{204}Pb = 15.322-15.596$ and $^{208}Pb / ^{204}Pb = 37.112-38.286$. It should be pointed out that: in terms of rocks, the tholeiitic and alkali — tholeiitic associations exhibit different degrees of enrichment, with the former being slighly higher in degree of enrichment; in terms of time, Eogene rocks are slightly enriched, Miocene rocks are relatively depleted and the isotopic values of the Quaternary approach the undifferented mantle values; in terms of space, the degree of depletion decreases from west to east, with the coastal area of eastern China showing a relatively high degree of enrichment. The quasi — primitive mantle type is identified from the Sr—Nd—Pb integrated indices and an overall evaluation of the time and space. So we should realize that the isotopic composition is characterized by 4—dimensional inhomogeneity in space and time and is changing constantly.

Isotope geochemistry of recycled enriched mantle—type volcanic associations   The typical areas of this type of volcanic rocks in China are Sanshui of Guangdong, the Changbai Mountains of Jilin and Tengchong of Yunnan. They are mainly related to paleoplate subduction and collision. These processes resulted in recirculation of crustal materials, particularly marine crustal materials, which were introduced into the wedge — shaped mantle of the continental lithosphere above the subduction zone; thus the original composition of the mantle of the continental lithosphere was modified and had the feature of enrichment in radioactive isotopes to different degrees. The associations of this type basalt are rather complex, including the trachybasaltic, basalt – trachyandesitic, trachyandesitic and trachytic association, the basalt – andesitic, andesitic, trachyandesitic and trachydacitic association and the alkali basaltic, tholeiitic, trachytic, tephritic and alkali rhyolitic associations. The SiO_2 content is 45.30 – 72.87%.

The common feature of this type of volcanic association is that the magma source region is the enriched mantle generated by recycling of crustal materials. The features of the Sr–Nd–Pb isotope system have a very wide variation range: $^{87}Sr/^{86}Sr = 0.704950 - 0.710980$, $^{143}Nd/^{144}Nd = 0.512503 - 0.512863$, $^{206}Pb/^{204}Pb = 17.422 - 18.685$, $^{207}Pb/^{204}Pb = 15.438 - 15.601$ and $^{208}Pb/^{204}Pb = 37.578 - 38.753$. In a certain sence, this is a binary mantle or a modified mantle resulting from the introduction of the isotopic composition representing crustal materials.

Isotope geochemistry of mixed mantle–type volcanic rock associations This type of mantle region is distributed in part areas of Jiangsu, Shandong and Anhui provinces and in northern part of Zhejiang province, concentrated in the middle – lower Yangtze region, i.e. the southeastern part of the North China block and at the northeastern end of the Yangtze block, in areas that underwent convergence and divergence many times in the geological history and interaction and in areas that underwent interaction between the lower crust and upper mantle materials due to repeated strike – slip compression and pulling – apart of the Tanlu fault system. The Cenozoic basalt associations in these areas are also rather complex. They show relatively wide variation in petrological and geochemical features. Eogene basalts are dominated by the subalkaline series basalt; Neogene basalts, the alkaline series; Quaternary basalts, the alkaline and peralkaline series with ultrabasic character. The tholeiitic series is represented by the olivine tholeiitic, tholeiitic and quartz tholeiitic association; the alkaline series is marked by the alkali olivine basaltic, basanitic, phonolite – basanitic and olivine nephelinitic association. The SiO_2 content varies between 42.68 and 53.66.

The compilation of the map has been carried out by the Institute of Geology of the MGMR with Profs. Li Zhaonai and Wang Bixiang as the chief editors. Bureaus of Geology and Mineral Resources and regional geological survey parties of various provinces, municipalities and autonomous regions specially assigned persons for the job (for details see the appended table of the explanatory notes). They shared out the work and cooperated with one another according to the unified requirements and technical standards. The members of the Map Compilation Group were Li Zhaonai, Wang Bixiang, Sheng Ruxiang, Wu Cailai and Shi Rendeng. The explanatory notes were written by Li Zhaonai. The computer drawing of the map was accomplished by Ma Guorui, Wang Liya, Su Lingfen and Guo Xingming of the Data Mapping Center of the Geological Survey of the Bureau of Geology and Mineral Resources of Hebei Province. The responsible editors for the publication were Tang Hanzhang and Tan Huijing. The terminology of related igneous rock groups in the map is based on the classification scheme (Le Bas, 1986) recommended by the International Union of Geological Sciences (IUGS) Subcommission on the Systematics of Igneous Rocks, while the nomenclature of the igneous rocks that are not concerned by the IUGS Subcommission but have important significance to the

Chinese continent is based on the classification scheme (Li Zhaonai et al., 1984) recommended by the Commission on Petrology of the Geological Society of China. For the terms of regional geology concerned in the map, in principle the terminology of the book "An Overview of Regional Geology of China" (Cheng, 1995) was used for reference. For the tectonic terms, we made the widest possible use of the conventional terms that are generally accepted in the world.

REFERENCES

1. Bureau of Geology and Mineral Resources of Anhui Province. *Regional geology of Anhui Province*, Geological Publishing House, Beijing (1984).

2. Cheng Yuqi (Ed). *The Introduction of Regional Geology of China*, Geological Publishing House, Beijing (1994).

3. Bureau of Geology and Mineral Resources of Fujian Province. *Regional geology of Fujian Province*, Geological Publishing House, Beijing (1985).

4. Bureau of Geology and Mineral Resources of Fujian Province. *Regional geology of Fujian Province*, Geological Publishing House, Beijing (1992).

5. Bureau of Geology and Mineral Resources of Gansu Province. *Regional geology of Gansu Province*, Geological Publishing House, Beijing (1991).

6. Bureau of Geology and Mineral Resources of Guangdong Province. *Regional geology of Guangdong Province*, Geological Publishing House, Beijing (1988).

7. Bureau of Geology and Mineral Resources of Guangxi Zhuang Autonomous Region. *Regional geology of Guangxi Zhuang Autonomous Region*, Geological Publishing House, Beijing (1985).

8. Bureau of Geology and Mineral Resources of Guizhou Province. *Regional geology of Guizhou Province*, Geological Publishing House, Beijing (1987).

9. Bureau of Geology and Mineral Resources of Hebei Province. *Regional geology of Hebei Province*, Geological Publishing House, Beijing (1989).

10. Bureau of Geology and Mineral Resources of Henan Province. *Regional geology of Henan Province*. Geological Publishing House, Beijing (1989).

11. Bureau of Geology and Mineral Resources of Heilongjiang Province. *Regional geology of Heilong Jang Province*, Geological Publishing House, Beijing (1993).

12. Hong Dawei, Guo Wenqi, Li Gejing, Kang Wei, and Xu Haiming. *The petrology of the belt of miarolitic granites in the southeast coast of Fujian Province and their generation*, Beijing Science and Technology Press, Beijing (1987).

13. Bureau of Geology and Mineral Resources of Hubei Province. *Regional geology of Hubei Province*, Geological Publishing House, Beijing (1990).

14. Bureau of Geology and Mineral Resources of Hunan Province. *Regional geology of Hunan Province*, Geological Publishing House, Beijing (1988).

15. Bureau of Geology and Mineral Resources of Jiangsu Province. *Regional geology of Jiangsu Province*, Geological Publishing House, Beijing (1984).

16. Bureau of Geology and Mineral Resources of Jiangxi Province. *Regional geology of Jiangxi Province*, Geological Publishing House, Beijing (1984).

17. Bureau of Geology and Mineral Resources of Jilin Province. *Regional geology of Jilin Province*, Geological Publishing House, Beijing (1989).

18. Bureau of Geology and Mineral Resources of Liaoning Province. *Regional geology of Liaoning Province*, Geological Publishing House, Beijing (1989).

19. Li Zhaonai and Wang Bixiang (Eds), *Volcanic rocks, volcanism and metallogenesis*, Geological Publishing House, Beijing (1993).

20. Li Zhaonai, Geological–geochemical types of Mesozoic rocks in eastern China, *Dixue Yanju*. Geological Publishing House, Beijing 26, 122–124 (1995).

21. Li Zhaonai, Magmatism and its deep–seated process in eastern China, *Progress in geology of China (1993–1996)– paper to 30th IGC*, Chian Ocean Pree, Beijing, 138–141 (1996).

22. Liu Ruoxin (Ed), *Geochemistry and geochronology of Cenozoic volcanic rocks of China*, Seismological Press, Beijing (1992).

23. Bureau of Geology and Mineral Resources of Nei Mongol Autonomous Region. *Regional geology of Nei Mongol (Inner Mongolia) Autonomous Region*, Geological Publishing House, Beijing (1991).

24. Bureau of Geology and Mineral Resources of Ningxia Hui Autonomous Region. *Regional geology of Ningxia Hui Autonomous Region*, Geological Publishing House, Beijing (1990).

25. Bureau of Geology and Mineral Resources of Qinghai Province. *Regional geology of Qinghai Province*, Geological Publishing House, Beijing (1991).

26. Ren Jishun and Chen Tingyu, et al. *Tectonic evolution and metallogenesis of the continent of eastern China and its neighbouring region*, Science Press, Beijing (1992).

27. Bureau of Geology and Mineral Resources of Shaanxi Province. *Regional geology of Shaanxi Province*, Geological Publishing House, Beijing (1989).

28. Bureau of Geology and Mineral Resources of Shandong Province. *Regional geology of Shandong Province*, Geological Publishing House, Beijing (1991).

29. Bureau of Geology and Mineral Resources of Shanxi Province. *Regional geology of Shanxi Province*, Geological Publishing House, Beijing (1989).

30. Bureau of Geology and Mineral Resources of Sichuan Province. *Regional geology of Sichuan Province*, Geological Publishing House, Beijing (1991).

31. Bureau of Geology and Mineral Resources of Shanghai Municipality. *Regional geology of Shanghai Municipality*, Geological Publishing House, Beijing (1990).

32. Tan Dongjuan and Lin Jingqian. *Mesozoic potassic magma province in north China platform*, Geological Publishing House, Beijing (1994).

33. Weng Shijie and Kong Qingshou et al. *Late Mesozoic volcanism of Zhejiang, Fujian, Jiangxi and Guangdong*, Geological Publishing House, Beijing (1987).

34. Bureau of Geology and Mineral Resources of Xizang Autonomous Region. *Regional geology of Xizang (Tibet) Autonomous Region*, Geological Publishing House, Beijing (1993).

35. Bureau of Geology and Mineral Resources of Xinjiang Uygur Autonomous Region. *Regional geology of Xinjiang Uygur Autonomous Region*, Geological Publishing House, Beijing (1993).

36. Xu Keqin and Tu Guangchi (Eds). Geology of granites and their relationship to metallogenesis, Jiangsu Press of Science and Technology, Nanjing (1984).

37. Xu Weliang, Chi Xiaoguo et al. *Mesozoic dioritic rocks and deep–seated inclusions in central North China Platform*, Geological Publishing House, Beijing (1994).

38. Bureau of Geology and Mineral Resources of Yunnan Province. *Regional geology of Yunnan Province*, Geological Publishing House, Beijing (1990).

39. Zhang Dequan and Sun Guiying. *Granites of eastern China*, China University of Geosciences Press, Wuhan (1988).

40. Zhang Ligang. *Block–geology of eastern Asia lithosphere*, Science Press, Beijing, 231–235 (1995).

41. Bureau of Geology and Mineral Resources of Zhejiang Province. *Regional geology of Zhejiang Province*, Geological Publishing House, Beijing (1989).

42. Zhou Xunruo and Wu Kelong et al. *I and A–type granites of Zhangzhou*, Science Press, Beijing (1994).

Proc . 30th Int'l. Geol. Congr. , Part 15, pp. 87 – 100
Li et al. (Eds)

Carbon Isotopic Compssition and Gensis of Diamond

Liu Guanliang[1], Han Youke[1], Zhai Lina[1], Miao Qing[2] Che Feng[2]

1. *Yichang Institute of Geology and Mineral Resources, CAGS P. C: 443003, No. 21 Gangyao Road Yichang City (P. O. Box 502) P. R. China.*

2. *6th Geological Brigade of the Liaoning Bureau of Geology and mineral Resources P. C: 116200, Pulandin city Liaoning province. P. R. China*

Abstract

The carbon isotopic composition of more than 230 diamond grains from three main diamond mines of kimberlite and lamproite and a placer deposits in China were investigated. The $\delta^{13}C$ values range from $-26.06\%_0$ to $+1.5\%_c$ with variation of about $28.00\%_0$. The obtained data indicates that no definite relationship occurred between the crystallization habit of diamond and their carbon isotopic composition but the colour and crystal type of diamond are certainly related to. The $\delta^{13}C$ of diamonds from different mines or different localities has their own feature. The data also show that the large difference in $\delta^{13}C$ valus is not attributed to the heterogeneity of the mantle while the definite effect are from the Archaean-Proterozoic subduction and recycling of the crustal carbon joining into the upper mantle. This paper proposes a "two suites and three stages" evolutionary model for explaining reasonably the large difference in $\delta^{13}C$ in diamonds. No carbon isotope differentiation occurred in the C-O-H closed system after the formation of kimberlite or lamproite magma. This new idea is of important significance not only to searching and prospecting of diamond but also universal to the inverstigation of the thickening mechanism of the south rim of Yangtze Craton and other areas in China.

Keywords: Diamond, Carbon Isotopic Composition.

I . INTRODUCTION

Investigations in initial stage indicated that the $\delta^{13}C$ values of diamond range from $-3.2 \sim -6.9\%_0$[3][32][30]with an average of $-5.8\%_0$. Afterwards, however, 4 carbonado grains from Aberyach, North Yakutia give $\delta^{13}C$ values varying from $-21.4\%_0$ to $-22.2\%_0$,[8] and a diamond grain from "Peace" pipe yields a $\delta^{13}C$ value of less than $-32.3\%_0$[9]. Obviously, carbon isotopic composition of diamond is much more complicated than that imagined.

More than 2,000 diamond grains have been studied so far [6,5],showing the $\delta^{13}C$ values ranged from of -34.5 to $+2\%_0$. However various diamond genetic models proposed by differeat authors are difficult to interpret such a large carbon isotopic difference. This paper proposes a "two suites and three stages" evolutionary model for interpretation such a large difference in $\delta^{13}C$ values of diamouds based on the data of carbon isotope determined on more than 230 diamond grains from 4 diamond mines from different localties in China.

Diamond samples used for carbon isotope analysis were well studied on the crystal form, colour, deformation and color dots of the crystals and their infrared spectra, ultraviolet absorption spectra and paramagnetic resonance spectra in order to identify the type and occurrence of mixed nitrogen.

Carbon isotopic compositions ($\delta^{13}C_{PDB}$ values, ‰) of all of samples are determined from CO_2 obtained by means of vacuum oxidation. Inclusion-bearing diamonds are determined after extraction of inclusions. Isotopic analyses were undertaken on a mass spectrometer MAT—251, with a standard deviation of ±0.1~0.2‰, in the Isotope Laboratory, Yichang Institute of Geology and Mineral Resoarces, MGMR, China.

II . BASIC FEATURES OF CARBON ISOTOPIC COMPOSITION OF DIAMOND

Carbon isotopic compositions for more than 230 diamond grains from kimberlite veins and pipes in Fuxian and Mengyin and from lamproite in Zhenyuan and from Dingjiagang-Taoyuan placer diamond deposits are shown in (Fig. 1). It shows that the $\delta^{13}C$ values for diamond in China vary from —26.06‰ to +1.52‰. However, most of the $\delta^{13}C$ values concentrated in between —9‰ and —2‰ with a peak value of —5~—4‰, forming a distribution pattern gently dipping to light carbon isotope. For

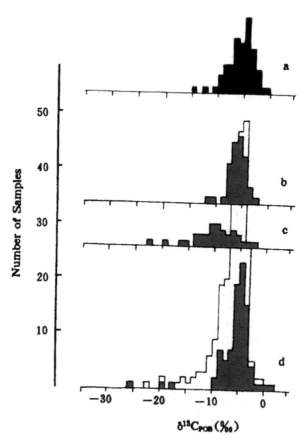

Figure 1.   Histogram of carbon isotopic composition of diamonds from different mines in China

a — Diamonds in kimberlite of Fuxian mine, Liaoning Province; b — Diamonds in kimberlites of Menying mine, Shandong Province; c — Diamonds in lamproites of Maping, Guizhou Province; d—Diamonds in Yuanshui diamond placer mines, Hunan Province.

details, The δ^{13}C values of diamonds from kimberlite in Fuxian, Liaoning vary from $-14.71‰$ to $-0.27‰$, with a peak value of $-5\sim-4‰$ (Fig. 1. a); diamonds from Mengyin mine yield δ^{13}C values varying from $-11.75‰$ to $-2.81‰$, especially, from $-8‰$ to $-3‰$ with a peak value of $-5\sim-6‰$ (Fig. 1. b), showing about $1‰$ isotope shift towards the light carbon isotope compared with those from Fuxian; diamonds from Zhenyuan mine give a δ^{13}C values of $-22.15\sim-2.53‰$, concontrated in -11 $\sim-5‰$ with a peak value of about $-10‰$ (Fig. 1. c). Which is characterized by more isotope shift towards the light carbon isotope in comparison with those from Fuxian and Mengyin; the δ^{13}C values of diamonds from Dingjiagang-Taoyuan placer diamond deposits range from $-26.06‰$ to $+1.52‰$, predominantly (more than 90%) from $-9‰$ to $2‰$ with a peak value of about $-5‰$ (Fig. 1. d). It should be pointed out that placer diamonds in Yuanshui River basin come from denuded primary diamond deposits. Therefore, to carry out a research for carbon isotopic composition of diamond is of importance to further prospecting and searching for diamond deposits. Summarizing up the above-mentioned data it is concluded that:

1. Carbon isotopic compositions of diamonds from different mines are different;
2. Diamond from North China Craton differs from Yangtze Craton in carbon isotopic composition probably indicating that upper mantle lithosphere underneath Yangtze Craton underwent a complicated evolutionary process during crystallization of diamond, while that underneath north China Craton was in a higher evolutionary stage during crystallization of diamond; 3. Probably, the carbon isotopic composition of kimberite type diamond largely differs from lamproite type diamond. The δ^{13}C value of diamond from Zhenyuan lamproite with the peak value of $-11\sim-10‰$ is similar to that from Arggle lamproite, west Australia[15]; 4. The peak value of δ^{13}C values of diamonds from placer deposits in Yangtze Craton is similar to that of diamonds from north China Craton, which indicates that not only lamproite type but also kimberite type diamond deposits might be formed in Yangtze Craton, China.

All of these indicate that carbon isotopic compositions for diamond in China are similar to those abroad, with the exception of slightly narrow distribution. The peak value of about $-4.6‰$ coincides with the initial carbon isotopic composition of mantle of $-9\sim-2‰$ with the average of $-5‰$ or $-4.6‰$ recognized by Meyer[21], Galimov[9]. It is probably characteristic of initial carbon isotopic composition of the Earth's mantle.

III. RELATIONSHIP OF COLOR, CRYSTAL FORM OF DIAMOND WITH THE CARBON ISOTOPIC COMPOSITION

The present paper carries out a research into color, crystal habit, plastic deformation, color dot of diamond grains prior to carbon isotope analyses and has the diamond grains classified, arranged in order of colorless, yellow, brown, green and black ones, monocrystal (octahedral, rhombic dodecahedral, hexoctahedral, cubic), combination (octa-dodecahedral, hexa-octahedral and so on), twin crystal, aggregate and the carbon isotopic compositions counted, respectively. The results indicate that crystal form of diamond bears no relation to the carbon isotopic composition, while most of colored diamond grains are in close relationship with light carbon isotopic composition. The former coincides with the conclusion reached by Deines, Gurney and Harris[5]when they study carbon isotopic compositions of diamond grains from Premier and Finsch mines, Southern Africa, while the latter is similar to the conclusion reached by Galimov[4] who points out that carbon isotopic composition of colored diamond is more complicated than that of colorless diamond and that diamond with light carbon isotopic composition (δ^{13}C value of less than -10%) predominates in colored diamond. It should be pointed out that diamond grains studied by Galimov occur in Laurasia land, while diamond grains studied by Deines et al occur in Gondwana land. The two different conclusions may be related to regional feature of mantle. The relationship between color and crystal habit of diamond in China and carbon isotopic composition is characterized by "intermediate type". It may be related to the regional feature of mantle characteristic of Chinese continent. On the other words, difference between geochemical provinces of mantle lithosphere is responsible for the relationship between color and crystal form of diamond and the carbon isotopic composition. However, factors coloring diamond such as accumulation of nitrogen and other impure elements, radiation damage and plastic deformation require further study.

IV. CARBON ISOTOPIC VARIATION WITHIN INDIVIDUAL DIAMONDS

Internal structure of individual diamonds is complicated. Assaying δ^{13}C values for different parts of individual diamond crystals cut by laser cutter, Swart et al[29] discover that carbon isotopic composition of the core of diammond crystals is lighter than that of the crust. Galimov[8] investigated variation in carbon isotopic composition within individual diamonds, and found that carbon isotopic composition of the crust of zonal diamond grains is characterized by small variation with a δ^{13}C value of $-5.9 \sim -8.1\%$, while that of the core is characterized by large variation with a δ^{13}C value of $-4.5 \sim -16.9\%$ and that carbon isotopic composition of uniform diamond grains only posses slight variation. According to the δ^{13}C values of different parts from two octahedral diamond crystals from "Udachnaya" pipe, Galimov[9] believes that carbon isotopic composition of different parts of the same diamond crystals is nonuniform which is attributed to multistage evolution of carbon reservoir in mantle.

Analyzing individual diamond grains for carbon isotopic composition by means of layered oxidation, the authors obtain the results[17,13] similar to the conclusion by Swart et al [29] but more complicated, describing as follows (Fig. 2):

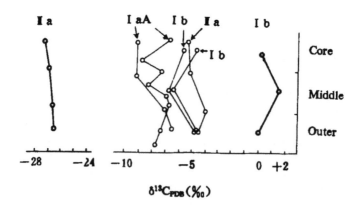

Figure 2.   Variation of δ^{13}C of the single-grain diamonds from outer to core

1. Carbon isotopic composition of the core is lighter than that of the crust, with a difference in $\delta^{13}C$ of 0.5~1.7‰;
2. Light carbon isotopic composition predominates in the core and crust, while heavy carbon isotope composition predomirates in the intermediate layer, with a difference in $\delta^{13}C$ of 1.5‰;
3. By contrast to the second, in some diamond graims heavy carbon isotopic composition predominates in the core and crust, while light carbon isotopic composition predominates in the intermediate layer, with a difference in $\delta^{13}C$ of 2.75‰;
4. Carbon isotopic composition for a zonal diamond crystal is characterized by alternation of light and heavy isotopes related to the zonal structure (Miao Qing, Liu Guanliang, Lu Qi et al[23]).

Variation in carbon isotopic composition of a diamond crystal during the growth may be attributed to either active carbon isotopic exchange reaction in carbon reservoir of mantle or heterogeneity of carbon isotopic composition of mantle. It should be noted although great difference in $\delta^{13}C$ between different diamond grains occurred, difference in $\delta^{13}C$ between different parts of the same diamond grain all ranges from 0.5‰ to 2.75‰. This indicates a little carbon isotopic fractionation during the crystallization process of diamond which couldnot be responsible for the large difference in $\delta^{13}C$ between diamond grains. In fact, such a large difference in $\delta^{13}C$ was formed in C-O-H system of mantle prior to crystallization of diamond.

V. RELATIONSHIP BETWEEN DIAMOND TYPES AND CARBON ISOTOPIC COMPOSITION

Natural diamond is formed in a unique geological setting. Diamond often contains impure elements such as H,B,Ni, N and so on among which N predoninates. Diamond may be classified on the basis of content and occurrence of N in it. Although, in practice, classification of diamond is much more complicated, according to Wen Luo et al, Zhu Hebal[37], Guo Jiugao and Tan Yimai et al[10], this paper classifies diamond into I aA, I aB, I b, I a, I b and combination types. However, there is no I b type diamond used for analysis in this study.
Fig. 3 Shows the relationship between diamond types and carbon isotopic compsition. It indicates that different diamond types are similar in distribution frequency of carbon isotopic compositions with apeak $\delta^{13}C$ value of $-4 \sim -6$‰. Howver, the $\delta^{13}C$ values from different types of diamond are different.
Relationship between diamond types from various mines and carbon isotopic composition is shown in Fig. 3. It indicates that distribution range of $\delta^{13}C$ values of I type diamond from 4 mines is wider than that of I type diamond. This implies that T,P and fO_2 of crystallization for I type diamond is much wider than those for I type diamond.

VI. DISCUSSION

More than 230 diamond grains in China yield $\delta^{13}C$ valus ranging from -26.06‰ to $+1.5$‰ and $\delta^{13}C$ values of more than 2,000 diamond grains from abroad are $-34.5 \sim +2.8$‰. Such large variation in $\delta^{13}C$ values of diamond is attributed to not only origin and evolution of C-O-H system in lithosphere but also genesis of diamond and geodynamics of the early Earth. Authors abroad believe that such large difference in $\delta^{13}C$

92

Liu Guangliang et al.

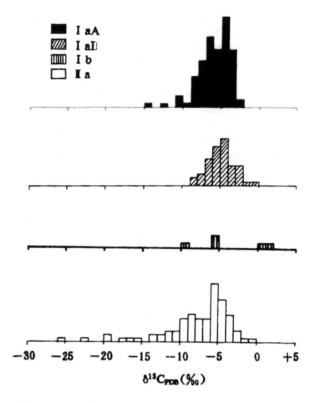

Figure 3.   Comparison of diamond types with carbon isotopic composition

values of diamond is a relection of difference in δ^{13}C values between different carbon
reservoirs related to heterogeneity of mantle (Deines[4]; Deines, Gurney, Harris[5]),
or a reflection of participation of crustal carbon in recycling of mantle, or both (Meyer
[21,22]). In addition, it is inferred that a part of diamond is formed from initial mantle
material while the other is a product of subduction (Ringwood Kesson [16, 24]).
However, Galimov[8,9] believes that formation of diamond is related to fractionation
other than subduetion.

At present, three ways can be chosen for interpretation for the large difference in δ^{13}C
values of diamond:
1. Carbon isotopic fractionation in C-O-H system of mantle
Carbon isotopic fractionation in C-O-H system of mantle is attributed to the isotopic
fractionation between total mantle carbon (C_M) and diamond carbon ($C_{diamond}$). Mantle
carbon occurs in crystalline phases such as diamond, graphite, carbonate, or as dissolved
carbon in solid solution or vapor phase. If mantle carbon crystallizes to form diamond
under condition that the C-O-H system is in equilibrium, no isotopic fractionation takes
place between mantle carbon and diamond carbon. Carbon isotopic fractionation occurs
only when P,T, f_{O_2} change (reservoir effect). That is to say, the degree of isotopic
exchange between various phases is controlled by the given total carbon and the changes of
other components in the system, i. e. δ^{13}C value of diamond depends on the degree of
carbon isotopic exchange between various phases (Deines[5]). Obviously, formation of

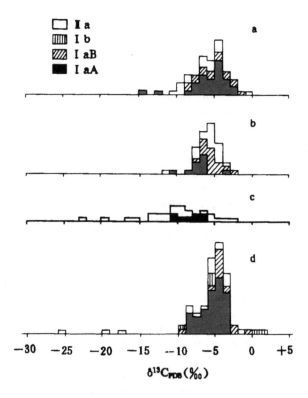

Figure 4.   Comparison of carbon isotopic composition with. diamond types from various mines
a—Fuxian mine; b—Menying mine; c—Maping; d—Yuanshui diamond placer mines.

diamond must result from isotopic fractionation between $C_{diamond}$ and C_M (C_M—total mantle carbon including CH_4, CO, C, CO_2 and so on). Therefore so long as suitable T, P, fO_2 conditions exist, mantle carbon can crystallize to form diamond. However, experimental data indicate a little fractionation occurred in this process. For example, an experiment of synthetic diamond shows that $\delta^{13}C$ values of graphite, diamond and residual graphite are $-25.80\%_0$, $-25.82\%_0$ and $-25.84\%_0$, respectively. Galimov's (1978) experiment shows isotopic shift of only $0.2-0.4\%_0$ towards light carbon isotope. Calculation indicates that under conditions of $1,000℃$, $\triangle_{(Diamond-Graphite)} = 0.4\%_0$, $\triangle_{(Diamond-CO_2)} = -4.4\%_0$ and $\triangle_{(CH_4-CO_2)} = -5.8\%_0$ (Galimov[8], Deines[4]). It is thus clear that carbon isotopic fractionation between solid phase, molten phase and vapor phase is very small, with a variation in $\delta^{13}C$ of less than $5-6\%_0$. And this fractionation will become negligible under high temperature conditions.

Carbon isotopic change during crystallization of diamond in C-O-H system is of significance to study of isotopic fractionation between $C_{Diamond}$ and C_M. The studies of Swart[29], Galimov[8,9], Han Youke[13], Liu Guanliang et al[17] indicate that carbon isotopic fractionation in the mantle C-O-H system in crystalline process of individual diamond grains is negligible with a variation in $\delta^{13}C$ of $0.5-2.75\%_0$. Morever, there are various carbon isotopic exchanges, for example, $^{12}CO_2 + ^{13}CO = ^{13}CO_2 + ^{12}CO$, $^{12}CO_2 + ^{13}CH_4 = ^{13}CO_2 + ^{12}CH_4$ and so forth in the C-O-H system. Crystallization of every diamond grain is limited to a given carbon isotopic range. It is evident that

carbon isotopic fractionation in the close mantle C-O-H system can not account for larg
range of $\delta^{13}C$ values of diamond.

Haggerty[11]points out that diamond might be formed from carbon in simple substance.
He believes that this sort of carbon, similar to plasma resulted from splitting of vapo
carbon on the interface between lithosphere and asthenosphere is an ideal carbon source o
microdiamond. If this carbon in simple substance is similar to atomic carbon by Frennd e
al (1980), isotopic fractionation of this atomic carbon can increase the light carbon
isotope range. But it can not result in the whole range of $\delta^{13}C$ values of diamond.

2. *Heterogeneity of mantle carbon source*

Heterogeneity of mantle is generally recognized. However, the question is what the caus
for this heterogeneity is? And how ever has the heterogeneity of mantle an effect or
carbon isotopic composition of diamond?

3. *Participation of crustal carbon in recycling of upper mantle*

It is well known that the average $\delta^{13}C$ value for crustal carbon is $-3 \sim -8\%_0$, while δ^{1}
C values for crustal fluid are comparable to those for diamond. For example, $\delta^{13}C$ valu
for endogenic CH_4 is $-2 \sim -35\%_0$, for CH_4 in carbonate film caught by garnet-
harzburgite in kimberite from Mengyin mine is $-21.62\%_0$ and for CO_2 is $-6.37\%_0$.
These indicate that participation of organic carbon such as CH_4 has a decisive effect on the
large difference of $\delta^{13}C$ values of diamond.

Mechanism of participation of crustal carbon in recycling of upper mantle is generally
attributed to subduction. If we relate the participation of crustal carbon in recycling of
upper mantle to formation of diamond, subduction must happen a long while ago.
Subcalcic garnet inclusion in diamond from Finsch and Kimberley, Southern Africa gives
an isotopic age of 3.3Ga, while the age of corresponding kimberlite intrusion is 150—
90Ma (Richardson[25]). Diamonds from Premier and Argyle mines are formed in
Proterozoic (Richardson[24]). Thus, a discussion on carbon isotopic composition of
diamond must back to carbon isotopic change in lithosphere during the Archean-
Proterozoic.

Inclusions in diamond are rich in Cr, Mg, LREE (Richardson, 1984) and OH^-(Wang
Alian, Wang Wuyi and Zhang Andi[31]). In recent years, Chen Feng[2] found
inclusions of KCl and NaCl in diamond, while the authors of this study found NaCl-
natural chromium-coesite inclusions, nepheline inclusions and chromite ($>60\%$ Cr_2O_3)-
moissanite-coesite inclusion occurred in diamonds. It seems that diamond contains not
only inclusions in peridotite or eclogite association but also composite inclusions formed
under a "chaos" condition. These suggest that diamond is formed in both depleted and
enriched mantle. Haggerty et al[12] found garnet with exsolution texture in xenolith in
kimberite in Jagersfontein, Southern Africa in 1990, while Gurney found Si-rich garnet
inclusions in diamonds from Southern Africa and Brazil in 1985. These xenoliths and
diamonds are considered as products in the transition zone with a depth of 410—660km.
Generation of diamond is much more complicated in comparison with that imagined.
Therefore, to investigate carbon isotopic composition of diamond should take vertieal and
lateral features of the whole upper mantle during formation of diamond into account.

Genetic models for diamond such as multiply constrained model (Haggerty[11]), model
of composition and structure of the Kaapvaal lithosphere, Southern Africa (Royd and
Mertzmans[27]; Helmstaedt and Gurney [14]), slab-mantle interaction model
(Ringwood[26][24]) have been established. Wyllie[33,34] carries out a research into
structure of lithosphere keel and subduction underneath continents and generation of
diamond. However, these models and researches donot deal with carbon isotopic
composition of diamond.

Existing information indicates that there are both in north China and Yangtze Cratons subduction, spreading and accretion of continental basin, extraction of basalt and komatiite taking place during Archean-Proterozoic. These events differ from those in other cratons in the world, by scope and the degree of depletion of mantle. In comparison with Yangtze Craton, north China Craton is characterized by higher evolutionary stage, higher homogenization and intense depletion of mantle which are confirmed by no diamond with light Carbon isotopic composition and also by initial Nd and Sr isotopic ratios. It is difficult to reconstruct the lithosphere keel of north China Craton because of intense disintegration since the Mesozoic-Cenozoic, probably, rifting during the Proterozoic. Therefore, searching the cause for the large difference in carbon isotopic composition of diamond in China must concentnate into the Yangtze Craton. So, this paper will deal with the relationship between the carbon isotopic composition of diamond and lithosphere evolution in Yangtze Craton only.

Obtained isotopic age of Yangtze Craton is younger than that of north China Craton. Age of the Kongling Group is $2,855 \pm 15$Ma (Liu Guanliang[32]). Upper Archean Sanyang Group with an age of $2.4-2.9$Ga and Middle Archean Longtouping Formation with an age of more than 2.9Ga are established (Zheng Weizhao, Liu Guanliang and Wang Xiongwu[36]). In the meanwhile, Ma Daquan (1991) obtains an isotopic age of $2.8-2.9$Ga for the Kongling Group. These suggest that Yangtze Craton may be compared with north China Craton.

Development and evolution of continental crust of Yangtze Craton in Archean have not been still clear, but subduction during the Early-Late Proterozoic is undoubted. Ultrabasic rock (ophiolite by Guo lingzhi et al, 1987) and komatiite (Mao Jingwen[20]) in the Sibao Group in Jiangnan island arc, basaltic Komatiite in the Matiyi Formation in Yiyang (Xiao Xidi[35]) and ophiolite in south Anhui and northeast Jianqxi (Bai Wenji[1], Zhou Xinming[18],) and ultrabasic-basic rock in the Banxi Group in west Hunan show multiphase subduction, collision orogeny and spreading in Yangtze Craton during the Proterozoic. These ultrabasic-basic masses yield Sm-Nd and Rb-Sr age values varying from 2,614Ma to 890Ma. Subduction is of importance to generation and evolution of mantle lithosphere and generation of diamond. At the same time, crustal material is bound to take part in recirculation of upper mantle and to enhance vitality of enrichment and metasomatism of the lithosphere. Subduction occurred in east Qinling subduction zone, north Yangtze Craton during the Middle-Late Proterozoic, too, forming an ultramafic-basic rock-andesite-acid volcanic rock association. Dahongshan lamproite zone with a length of more than 100km is exposed in the northern margin of Yangtze Craton (Liu Guanliang, Wang Xionguw and Lü Xuemiao [19]) and Changsha-Guiyang lamproite zone consisting of Ningxiang, Zhengyuan and Majiang masses is exposed in the southern margin of Yangtze Craton (Fig. 5). From this it is inferred that a potential diamond area is located in Yangtze Craton.

The geophysical transection information indicates that the present interface between lithosphere and asthenosphere underneath Yangtze Craton is about 150km deep, while the lithosphere keel with a depth of more than 280km is to the southeast of Jiangnan Island arc. A study on inclusion mineral association in diamond indicates that diamond from Yangtze Craton is formed at a depth of more than 230km. That is to say, there was a giant lithosphere keel underneath Yangtze Craton when diamond was formed.

Generation and evolution of the lithosphere keel underneath Yangtze Craton may be divided into the following three stages shown in (Fig. 6.)

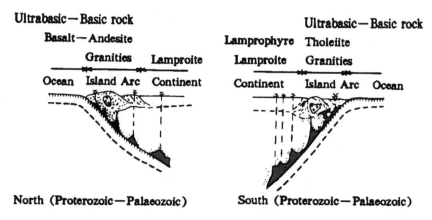

Figure 5.   Subduction related to magma suites in the middle part of Yangtze Craton
             (after Liu Guanliang and Xu Tao,1982).

1. Subduction stage. The studies on calc-alkaline volcanic suite in the marginal island arc and ultrabasic rock in the subduction zones indicate subduction of oceanic plates occurred on both (south and north) sides of Yangtze Craton. Subdnction of oceanic plates into mantle at a depth of 200km underneath the continent, forming a composite structure consisting of continental crust, lithosphere and astherosphere and oceanic crust, lithosphere and astheosphere (Fig. 6a).

2. Ultrametamorphism- metasomatism stage. Difference in chemical component and varation in temperature, pressure and oxygen fugacity between lithosphere and asthenospbere and participation of volatile matter (plume) from asthenosphere are bound to result in widespread ultrametamorphism and metasmatism (Fig. 6b). Mixture of those from crust with C- O- H from asthenosphere enhances chemical process between active components, in the mean while, they continuously permeat and replace the depleted lithosphere. In addition, crustal carbon takes part in recirculation of upper mantle. Diamond should generate at this moment. Basalt from oceanic crust is metamorphosed into diamond-bearing eclogite in depth, while serpentinite is metamorphosed into depleted harzburgite and dunite. Harzburgite and diamond which are charaeterized by plastic deformation are formed under conditions of mantle convection. It is well known that diamond crystallizes under conditions of CH_4 oxidized and CO_2 reduced. This is easy to come true when crustal material mix with mantle material and oxygen fugacity changes. This may be verified by the existence of inclusions of KCl, NaCl, nepheline and composite inclusion of chromite-coesite-moissanite in diamond or by frequent variation in carbon isotopic composition during crystallization process of diamond. This is the reason why difference in $\delta^{13}C$ values of diamond is so large.

3. Homogenization. Ascending mantle current continuously homogenizes during ultrametamorphism and metasomatism, resulting in a new and thickening lithosphere keel in order to be in equilibrium under new conditions (fig. 6c).

$\delta^{13}C$ value in the new lithosphere should be about $-5\%_o$. A large amount of diamond grains crystallize at this time. Continental lithosphere is still a geochemical barrier for the thickening part of the new lithosphere. Therefore, under suitable conditions, diamond will be crystallized according to Haggerty's (1986) model. Partial melting of the

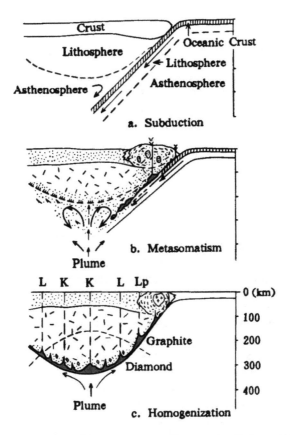

Figure 6.   The evolution model of ancient lithosphere keel of the middle part of Yangtze Craton. K—Kimberlite, L—Lamproite, Lp—Lamprophyre.

lithosphere during metasomatism and degassification of mantle should result in primitive magmas of kimberlite, lamproite and nephelinite into which incompatible elements and high field strength elements migrated. After $n \times 100Ma - n \times 1,000Ma$, these magmas with diamond from different source areas erupted, forming kimberite and lamproite.

Carbon isotopic fractionation in the closed system of kimberlitic or lamproitic magma is too weak to result in the large difference in $\delta^{13}C$ values of diamond, which has been verified by $\delta^{13}C$ values of cubic diamond containing inclusions of crustal material and magma melt glass.

The above-mentioned may be called "two suites and three stages" evolutionary model for the lithosphere keel. Subduction stage of the evolutionary model is of importance to the large difference in carbon isotopic composition of diamond because of participation of crustal carbon, especially, CH_4 in recirculation of mantle. In addition, the participation of crustal carbon accounts for $\delta^{13}C$ values of diamord in eclogite association and irregular variation in $\delta^{13}C$ values within invidual diamonds during the growth. This evolutionary model may be used for making an approch to mechanism of upper mantle lithosphere thickening (keel) in the southeastern margin of Yangtze Craton and underneath

Cathaysian Craton and is of theoretical significance to prospecting of primary diamond deposits in Yangtze Craton.

VIII. CONCLUSIONS

Above-mentioned study on carbon isotopic composition of diamond samples from 4 mines in North China and Yangtze Craton points to the following conclusions:

1. Existing $\delta^{13}C$ values of diamonds in China range from $-26.06‰$ to $+1.5‰$, with a concentrated range of $-9 \sim -2‰$ and the highest peak value of about $-5‰$.

2. Carbon isotopic composition of diamond inluences the color and crystal form of diamond. However, the $\delta^{13}C$ values of diamonds from different mines or localities have their own features.

3. Carbon isotopic composition of diamond is related to type of diamond. Most of the $\delta^{13}C$ values from I type diamonds are $-9 \sim -2‰$; the $\delta^{13}C$ values from I type diamonds are $-26.06 \sim 0‰$ with the peak value characterized by slight shift towards light isotope.

4. Diamonds from North China Craton differ from those from Yangtze Craton by carbon isotopic composition. Range of the $\delta^{13}C$ values of diamonds from Yangtze Craton is wider than that from north China Craton. This is a reflection of difference in genetic mechanism and evolutionary stage between them. The Yangtze Craton underwent more intensive subduction.

5. Variation in $\delta^{13}C$ values within single diamond crystal is irregular, being in contradiction to Rayleigh fractionation principle. It is a reflection of rapid change of carbon isotopic composition of the carbon source in mantle during diamond crystallization.

6. The large difference in $\delta^{13}C$ values of diamonds should be related to the carbon source in mantle during the Archean-Proterozoic. Carbon isotopic fractionation in the closed C-O-H system of kimberite or lamproite magma took place according to Rayleigh fractionation principle which should not result in large difference in $\delta^{13}C$ values.

7. "Two suites and three stages"-evolutionary model for the lithosphere keel during the Archean-Proterozoic may be used for explaining the large difference in $\delta^{13}C$ values of diamond. Participation of crustal carbon, especially, CH_4 in recirculation of upper mantle is of special significance.

8. " Two suites and three stages"-evolutionary model may be used for making an approach to mechanism of lithosphere thickening (for example, lithosphere thickening in the southeastern margin of Yangtze Craton).

Acknowledgement. The authors convey their thanks to Senior Engineer Zhang Peiyuan, MGMR, P. R. China, Senior Engineer Han Zhuguo, 6th Geological Brigade of the Liaoning Bureau of Geology and Mineral Resources, Senior Engineers Hu Shiyi, Hu Shijie, 7th Geological Brigade of the Shandong Bureau of Geology and Mineral Resources, Senior Engineer Ma Wenyun, 413th Geological Brigade of the Hunan Bureau of Geology and Mineral Resources, senior Engineers Luo Huiwen, and Ren Huaixiang, 101th Geological Brigade of the Guizhou Bureau of Geology and Mineral Resources for their support to this study. We thank senior research Fellows Riao Jiaguang, Ma Daquan, Zhang Zichao and Zhang Ligang for their useful suggestions on the manuscript.

This work was supported by the National Nature Science Foundation of China

REFERENCES
1. Bai Wenji, Gan Qigao, Yang Jingsui, Xing Fongming and Xu Xiang. Discovery of well — reserved ophiolite and its basical characters in southeastern margin of the Jiangnan Ancient continent. Acta Petrological et Mineralogica. 1, 289—299 (1986)
2. Chen Feng, Guo Jiugao, et al. , High—K and high—Cl inclusions in diamond and mantle metasomatism. Acta Mineralogical Sinica. 3. 193—198 (1992).
3. Craig, H. , The geochemistry of stable carbon isotopes. Geochimica et Cosmochimica. Acta, 3,53—92. (1953).
4. Deines, P. , The carbon isotopic composition of diamonds relationship to diamond shape, color, occurrence and vapor composition. Geochimica et cosmochimica Acta. 44, 943—961 (1980).
5. Deines, P. , Gurney, J.J. and Harris, J.W. , Associated chemical and carbon isotopic composition variation in diamond from Finsch and Premier Kimberlite. South Africa. Geochimica et cosmochimica Acta. 48, 325—342 (1984).
6. Galimov, E.M. , Problem of the origin of diamonds in the light of new duta in the carbon isotopic composition of diamonds. Abstr. VII Nat. Symp. Isot. Geochem. Moscow, 13. (1978).
7. Galimov, E.M. , Sobolev, N.V. , yefimova, E.S. , Shemanina, E.I. , and Maltsev, K.A. , Carbon isotopic composition of diamonds, containing mineral indusions, from placers of northern ural. Geokhimiya 9. 1363—1370 (1989).
8. Galimov, E.M. , The relation between formation conditions and variations in isotope composition of diamonds containing crystalline inclusions. Doklady Academii Nauk USSR, 249 (5). 1217—1220 (1979).
9. Galimov, E.M. , Isotope fractionation related to kimberlite magmatism and diamond formation. Geochimica et cosmochimica Acta. 55. 1687—1708 (1991).
10. Guo Jiugao, Tan Yimei et al. Type I b diamond in diamond placer mines, Huan province, Kexue Tongbao 31.4, 257—262 (1986).
11. Haggerty, S.E. , Diamond genesis in multiply constrained model, Nature, 320. 34—38 (1986).
12. Haggerty, S.E. , Petrochemistry of ultradeep (> 300km) and trasition zone xenoliths. Sixth IKC. extended abstracts, 214—216 (1995).
13. Han Youke and An Na, A stripping combustion method for the analysis of carbon isotopes of diamonds. Rock and Mineral Analysis, 7. 5 296—303 (1988).
14. Hlemstaedt, H. and Gurney, J.J. , Kimberlite of southern African, Are they related to subduction processes? 3th IKC. 1. 425—434 (1984).
15. Jaques, A.L. , Hall, A.E. , Sheraton, J.W. , Smith, C.B. , Sun, S.S. , Drew, R.M. , Foudoulis, C. , and Ellingsen, K. ,Composition of crystalline inclusions and C—isotopic.
16. Kesson, E.E. , and Ringwood, A.E. , Slab mantle interactions 1. Sheared and refertilised garnet peridotite xenoliths-samples of Wadati—Benioff zones? 2. The formation of diamonds, Chemical Geology, 70. 83—118. (1989).
17. Liu Guanliang and Wang Xiongwu, On the geological condition for the formation of type I diamond. Bulletin of the Yichang Institute of Geology and Mineral Resources. (CAGS), 14, 41—81 (1989).
18. Liu Guanliang. , New progression on age of kongling group, Reginal Geology of China 1. 93 (1987).
19. Liu Guanliang, Wang Xiongwu and Lu Xuemiao. , Dahongshan Lamproites. Geological publishing House. (1993).
20. Mao Jingwen,Zhang zongqing and Dong Baolin. A new Sm—Nd isotopic chronology

of SiBao group in southern margin of Yangtze Massif. Geological Review. 36. 1, 264—268 (1990).

21. Meyer, H. O. A. , Inclusions in diamond. In Mantle Xenoliths Ed. by P. H. Nixon. 501—523 (1987).

22. Meyer, H. O. A. , Genesis of diamond a mantle saga. American Mineralogist. 70. 3, 344—350 (1985).

23. Miao Qing, Liu Guanqiang, Lu Qi, Wang Xiongwu and Zheng Shu, Discovery of new complex inclusions in diamond and their genesis. Geological Science and Technology in formation. 10. 117—124 (1991).

24. Richardson, S. H. , Latter—day origin of diamond of eclogite paragensis. Nature. 332. 623—626 (1986).

25. Richardson, S. H. , Gurney, J. J. , Erlank, A. J. and Harris, J. W. , Origin of diamonds in old enriched mantle. Nature. 301, 198—202 (1986).

26. Ringwood, A. E. , Slab mantle in teractions 3. petrogenesis of intraplate magmas and structure of the upper mantle. Geomical Geology, 82. 187—207 (1990).

27. Royd, F. R. and Mertzman, S. A. , Composition and structure of the Kaapvaal lithosphere. Southern Africa, The Geochemical society, special publication, 1. 13 —24.

28. Sobolev, N. V. , Galimov, E. M. , Ivnovskaya, I. N. and Yefimova, E. S. , Carbon isotope, composition of diamonds containing crystalline inclusions. Doklady Academii Nauk 249. 5 1217—1220 (1979).

29. Swart P. K. , Pillinger, C. T. , Milledge, H. J. , and Seal, M. , Carbon isotopic variation within individual diamonds. Nature, 303. 793—795 (1983).

30. Vinogradov, A. P. , Kropotova, O. T. , Orlov. Yu. L. and Grinenko, V. A. , Isotopic composition of diamond and carbonado crystals. Geokhimiya. 12. 1395— 1397 (1966).

31. Wang Wuyi, Guo Lihe, Wang Alian, Zhang Andi. A study of constitutional water in pyrope. Acta petrological Et Mineralogica. 11. 61—69 (1992).

32. Wickman, F. E. The cycle carbon and stable carbon isotopes. Geochimica et cosmochimica Acta, 9. 136—153 (1956).

33. Wyllie, P. I. The genesis of kimberlites and some low—SiO$_2$ high—alkali magmas. 4th IKC. Kimberlites and related rocks, 1. 603—615 (1989).

34. Wyllie, P. J. Metasomatism and fluid generation in mantle xenoliths. Ed. by P. H. Nixon. 609—612 (1987).

35. Xiao Xidi, On the basaltic komatiite and its formative environment in Yiyang, Hunan. Journal of Central — South Institute of Mining and Metallurgy. 38. 4. 106—113 (1983).

36. Zheng Weizhao, Liu Guanliang and Wang Xionguw. A new information of Archaean Eon for the Kongling Group in northern Huangling anticline. Hubei. Bulletin of the Yichang Institute of Geology and Mineral Resources (CAGS). 16. 96—107 (1991).

37. Zhu Hebao, Den Huaxing, Liu Hongji and Zhang Ziqian, The discovery of high content of Type I diamond in a kimberlite from chiua and its geology implications. Acta mineralogical Sinica. A. 283—289. (1982).

Proc. 30th Int'l. Geol. Congr., Part 15, pp. 101 – 108
Li et al. (Eds)
© VSP 1997

Vertical Chemical Variation of the Mesozoic Synuplift Plutonic Rocks in Dabieshan Orogen, Central China

MA CHANGQIAN, YANG KUNGUANG and XU CHANGHAI

Faculty of Earth Sciences, China University of Geosciences, Wuhan 430074, P.R.C.

Abstract

The Dabieshan orogen had undergone strong differential uplift-exhumation during the Mesozoic. The Beihuaiyang massif located on the north side of the Tongbai-Tongcheng Fault Zone (TFZ) has been of an exhumation magnitute of less than 5 km since 170 Ma. On the south side of the TFZ, the rocks of the East Dabie massif now at the surface were buried to a depth of approximately 18 km at 227 Ma, and plutonic emplacement of ~220-120 Ma was accompanied by uplift-exhumation of 8-10 km. There is a series of Mesozoic plutons exposed in East Dabie massif, which were emplaced synchronously with the rock uplift-exhumation of Dabieshan Mountains. This paper addresses the vertical chemical variation of the Mesozoic synuplift plutons at Dabieshan orogen. It is suggested that Fe_2O_3, TiO_2, P_2O_5, MgO, CaO, Al_2O_3 and CIPW-normative An are positively correlated with plutonic emplacement pressure; SiO_2, K_2O, Rb, Ba, Zr, Hf and CIPW-normative Or are negatively correlated with the pressure; and the vertical chemical variation should be mainly controlled by buoyancy and upward migration of pore-fluid material.

Keywords: granitoid rocks, vertical magma system, rock uplift-exhumation, Mesozoic, Dabieshan orogen

INTRODUCTION

Granitoid magma has played an essential role in the growth and differentiation of the continental crust. The redistribution of mass within the crust by the ascent of granitoid magma is an obvious example of this igneous connection [2]. However, less obvious but perhaps no less important is the vertical evolution of granitoid magma when it rose through the continental crust. The plutons that were emplaced synchronously with uplift of a mountain range offer the opportunity to understand the vertical magma system. The Mesozoic igneous belt exposed in the Dabieshan orogen, central China is a clear example of the synuplift vertical magma system. The intention of the paper is to reconstruct the Mesozoic vertical magma system from the Dabieshan orogen, on the basis of the studies of emplacement pressure-age relationship, and to discuss its vertical chemical variation.

REGIONAL GEOLOGIC SETTING

The Dabieshan orogen is tectonically located between the Sino-Korean Craton and the Yangtze Craton, and is considered to be the root belt of the Qinling orogen in central China, which has been the focus of intense geological studies for decades because of the discovery of coesite- and diamond-eclogites in the ultra-high pressure metamorphic zone[11,7,13]. It can be divided, in terms of the differences of stratigraphic, deformational, metamorphic and igneous features, into 5 massifs bounded by a series of fault zones: Beihuaiyang, Tongbai, Dawu, Dabie and Suiying massifs（Fig. 1）. The paper deals only with the Beihuaiyang, Dawu and Dabie massifs.

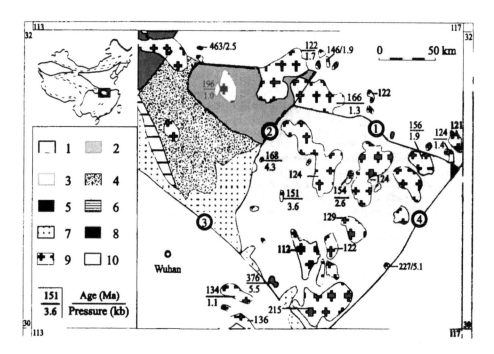

Figure 1. Geological sketch map of the Dabie Mountain region showing hornblende ^{40}Ar/^{39}Ar plateau ages (Ma) and total-aluminium geobarometry data (kb) of granitoids. 1. Beihuaiyang massif; 2. West Dabie massif; 3. East Dabie massif; 4. Dawu massif; 5. Tongbai massif; 6. Suiying massif; 7. Cretaceous - Cenozoic basin; 8. Paleozoic granitoids; 9. Mesozoic granitoids; and 10. Yangtze Craton. Fault zones: ① Tongbai-Tongcheng fault; ② Macheng-Shangcheng fault; ③ Xiangfan-Guangji fault; and ④ Tan-Lu fault.

The Beihuaiyang massif is composed of metasedimentary rocks developed from the middle-upper Proterozoic to Carboniferous. The Carboniferous rocks of paralic facies are exposed piecemeal in the massif. It has already been confirmed that there once occurred granitoid intrusions and mafic-felsic volcanic rocks with the Caledonian, Hercynian and Cretaceous ages [5]. At the eastern end of the massif, the Cretaceous alkaline intrusive rocks are also present. The massif had undergone metamorphism of greenschist facies, and as early as 1980's, it was confirmed to embrace the high-pressure metamorphic zone containing 3T-phengite [14]. The massif was dated by ^{40}Ar/^{39}Ar on amphibole in mafic metamorphic rocks, which yielded plateau age of 401 ± 4 Ma [4].

The Dawu massif mainly consists of a set of the middle-upper Proterozoic metasedimentary and volcanic rock series containing phosphorus, where metamorphism extends to greenschist facies. The oldest U-Pb zircon age is 2343 Ma [9]. The NW-striking glaucophane eclogite zone was developed in the massif, and blueschist was exposed at the south-east margin of the massif. The Cretaceous granitoids are identified .

The Dabie massif includes the Archean Dabie group (U-Pb zircon ages range from 2820-2413 Ma, [9]), the Lower-Middle Proterozoic Hong'an group and numerous granitoidplutons. The Dabie group is composed of supracrustal sequence, granitic gneiss and metabasic rocks with their corresponding percentages in the total area of the region being 38 %, 60% and 2% respectively. The supracrustal sequence is mainly made up of amphibole plagiogneiss, amphibolite, leucogranulitite and a small part of marble and magnet quartzite. The lower part of the Hong'an group is metasedimentary rocks with phosphorite deposits, and its upper part is metavolcanic rocks. The granitoid and dioritoid plutons ranging from the Neoproterozoic to Cretaceous are exposed. The Dabie massif can be separated into two sections: east and west by the Shangchang-Machang fault. The eastern section (*East Dabie massif*) is a well-known ultrahigh-pressure terrane containing coesite and diamond and have undergone the retrogressive metamorphism of amphibolite facies [15], locally up to granulite facies [9].There only occurred coesite eclogite in the western section (*West Dabie massif*), which have only undergone retrogressive metamorphism of epidote amphibolite facies.

The K-Ar ages of metamorphic biotite from the Dabieshan orogen show a region-wide variation [6]. The statistics indicate that 19 age data average 256 ± 85 Ma in the Beihuaiyang massif, 12 age data give an average value of 200 ± 44 Ma in the Dawu massif and the West Dabie massif, and 23 age data average 145 ± 55 Ma in the East Dabie massif. Since the closure temperature of Ar in biotite is $300° \pm 50$ ℃[8], the K-Ar biotite ages represent the cooling time of the rocks. Therefore, it can be supposed that the regional variation of K-Ar biotite ages should reflect the chronological relationship of rock uplift-exhumation between these massifs.

DETERMINATION OF SYNUPLIFT PLUTONIC ROCKS

Dabieshan plutonic rocks consist primarily of hornblende-biotite monzodiorite, quartz monzodiorite, quartz monzonite and granites (classification according to Streickeisen, [10] (Fig. 2); minor amounts of syenite and granodiorite are also present. Rock compositions are plotted in the calcalkaline field on an AFM diagram. They have an alkali lime index of ~59, moderately high amounts of K_2O (0.8~5.0 wt.%) and Al_2O_3 (12.8~19.5 wt.%), and metaluminous character (molar Al_2O_3 ($CaO+Na_2O +K_2O$) = 0.55-1.08).

Pluton emplacement pressures can be used to provide information, combined with geochronological data, about the rock uplift-exhumation magnitude [6], and pluton emplacement (or crystallization) pressures of the Dabieshan orogen can be obtained from the total-aluminium geobarometry of calcareous hornblende [1].

The Beihuaiyang and West Dabie Massifs

The nine samples from the Mafan intrusive rocks (462.7 ± 1.4 Ma, $^{40}Ar/^{39}Ar$ hornblende plateau age) in the Beihuaiyang massif yield an average pressure of 2.92 ± 0.18 kbars, and the Mesozoic quartz monzonites and granites give average pressure of 1.71 ± 0.05 kbars (Shangcheng pluton not far from the Mafan pluton, 146.0 ± 0.9 Ma,

^{40}Ar/^{39}Ar biotite plateau age) and 1.26 ± 0.29 kbars (the Shibei pluton, 166.4 ± 0.3Ma, ^{40}Ar/^{39}Ar hornblende plateau age). In addition, the average hornblende pressure of the Youzhahe quartz monzonite (196.2 ± 2.1 Ma, ^{40}Ar/^{39}Ar hornblende plateau age) in the west Dabie massif is 0.78 ± 0.14 kbars (Fig. 3). The data indicate that the uplift-exhumation of the Beihuaiyang and West Dabie massifs mainly occurred in the period between the Paleozoic and Jurassic, and the Mesozoic differential subsidence resulted in the formation of volcanic-sedimentary basins in the Beihuaiyang area.

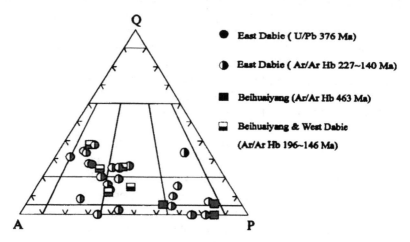

Figure 2. Modal composition of the Dabieshan plutonic rocks based on classification of Streickeisen [10].

In the Beihuaiyang, West Dabie and Dawu Massifs, a series of radial and ring granitic porphyry veins is distributed around the Shangcheng, Xinxian-Youzhahe, and Linshan intrusive bodies [6], cutting across the intrusive bodies. The granitic porphyry veins, often as wide as several meters to more than twenty meters and extending tens of kilometers, contain K-feldspar macrocrystals with the medium to coarse grain texture for its matrix. The mineralogy and chemistry of the porphyry veins are nearly the same from the east to the west, and some of their chemical features are also similar to those of the Xinxian and Laoshanzhai granitic intrusions. Two K-feldspar macrocrystals from the veins give the K-Ar ages of 95 and 92 Ma [6]. The veins did not occur in the East Dabie massif. These facts indicate that the Beihuaiyang, West Dabie and Dawu massifs were of same magmatic and uplift-exhumation history during the Mesozoic.

The East Dabie Massif

The strong rock uplift-exhumation of the East Dabie massif occurred in the Mesozoic. The hornblende crystallization pressure of the Hercynian Wuyueshan syenogranite (376 ± 26 Ma, U-Pb zircon method [3]) was 5.47 ± 0.18 kbars, far greater than that of the Mafan pluton in the Beihuaiyang massif. The Triassic Liujiawa gabbro-monzodioritic intrusion (227.3 ± 7.6 Ma, ^{40}Ar/^{39}Ar hornblende plateau age) has a higher hornblende pressure, 5.10 ± 0.56 kbars, and the Jurassic Xixiongwen (168 ± 17 Ma, U-

Pb zircon method, [3]) and Xiaozhai (151.2 ± 1.5 Ma, [40] Ar/ [39] Ar hornblende plateau age) intrusions have hornblende pressures of 4.33 ± 0.44 and 3.57 kbars, respectively. Thus, the rock uplift-exhumation extended for about 5~6 km from the Triassic to Jurassic period (Fig. 3).

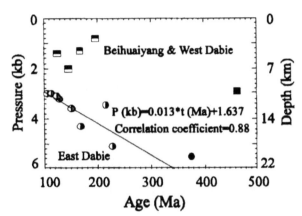

Figure 3.   Plot of emplacement ages(Ma) versus pressures(kbars) for the Dabieshan pluton.

VERTICAL CHEMICAL VARIATION OF THE PLUTONIC ROCKS

The Mesozoic plutonic rocks in the Dabieshan orogen, especially East Dabie massif, show considerable variation in composition with emplacement pressures. For example, Fe_2O_3, TiO_2, P_2O_5, MgO, CaO, Al_2O_3 and CIPW-normative An increase with the emplacement pressures, and SiO_2, K_2O, Rb, Ba, Zr, Hf and CIPW-normative Or are negatively correlated with the pressures (e.g., Figs. 4 -5), indicating that enrichment of high-temperature, heavier components in the lower part of vertical magma system, and enrichment of some low-temperature, lighter components and pore-fluid material in the upper part of the magma system.

It is noteworthy that the Beihuaiyang and East Dabie massifs show different composition-pressure relationship, and the Paleozoic granitoids in East Dabie drift off the correlation between the rock composition and emplacement pressure of the Mesozoic plutons. In addition, some co-denominator element ratio-ratio plots show straight lines related with magma mixing (Fig. 6).

A complication that could affect the correlation between the rock composition and emplacement pressure is the effect of pluton volume, emplacement mechanism and thermal relationship between pluton and its wall rocks. The upward height (h) of a spherical magma body can be given by

$$h = \sqrt{8Lr/3c\beta}$$

where L is a latent heat of magma crystallization, r is radius of magma body, c is specific heat of magma, and β is geothermal gradient [12]. It is apparent that, other things being

Ma Changqian et al.

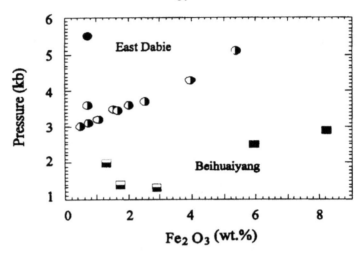

Figure 4. Illustration of variation of Fe2O3 (wt.%) with emplacement pressure (kbars).

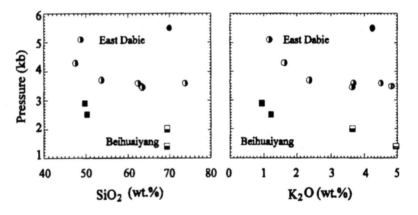

Figure 5. Illustration of variation of SiO2 and K2O (wt.%) with emplacement pressure (kbars).

equal, under larger geothermal gradient pluton would have smaller upward height, and that different emplacement depths between the Paleozoic and Mesozoic granitic plutons in the East Dabie, and between the Mesozoic plutons of Beihuaiyang and East Dabie massifs could be results of magma emplacement under different geothermal gradients.

This equation shows that magma with larger specific heat would rise at a shorter distance As hornblende, biotite and calcium-rich plagioclase have larger specific heat than potassium feldspar and quartz [16], the magmas with more dark-mineral and calcium-rich plagioclase components should be emplaced in the deeper crust level. It is suggested that the vertical composition variation of the Mesozoic plutons in the East Dabie massi should be mainly controlled by buoyancy , and the upward migration of potassium-rich pore-fluid material or autometasomatism had played an important role in the vertical variation of magma composition. The last interpretation is supported by some co-denminator element ratio-ratio plots (fig. 6).

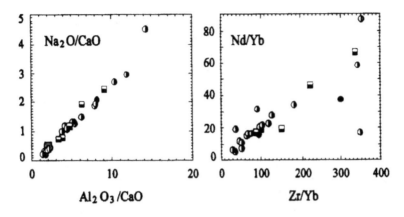

Figure 6. Plots of Al2O3/CaO versus Na2O/CaO and Zr/Yb versus Nd/Yb.

CONCLUSIONS

The Dabieshan orogen had undergone strong differential uplift-exhumation. The estimate of pluton emplacement pressures indicates that the rock uplift-exhumation magnitude in the Beihuaiyang massif was only about 10 km since the Caledonian, and that of the East Dabie massif was 20 km or so since the Hercynian.

The Mesozoic calc-alkaline plutons in East Dabie massif were emplaced synchronouslywith strong rock uplift-exhumation. The rocks in the East Dabie massif were buried at the depth of approximately 18 km at 227 Ma, and pluton emplacement of 220-150 Ma was accompanied by uplift-exhumation of 5~6 km.

The vertical chemical variation of the Mesozoic plutonic rocks could result from buoyancy and upward migration of potassium-rich pore-fluid material.

Acknowledgments

The study was supported by grants from Fok Yingtung Education Foundation and NSFC (No.49572100). We thank Prof. Wang Songshan for his help in $^{40}Ar/^{39}Ar$ dating, and Prof. Li Zhaonai for thorough review of the manuscript. Yan Yuqin aided us in sample preparation.

REFERENCES

1. M.C. Johnson and M.J. Rutherford. Experimental calibration of the aluminum-in-hornblende geobarometer with application to Long Valley Caldera (California) volcanic rock, *Geology*.17, 837-841(1989).
2. R. W. Kay, S. Mahlburg Kay and R. J. Arculus. Magma genesis and crustal processing. I n: *Continental Lower Crust*. D.M. Fountain, R.Arculus and R. W. Kay (Eds.). pp. 423-445. Elsevit, Amsterdam (1992).
3. S. Li and T. Wang. *Geochemistry of Granitoids in Tongbaishan-Dabieshan, Central China*. Press of China University of Geosciences, Wuhan (in Chinese with English abstract) (1989).

4. Z. Liu, B. Niu and J. Ren. Disintegration of the Xinyang group and its tectonic implication, *Geological Review.* 38, 293-301(in Chinese with English abstract) (1992).

5. C. Ma, K. Yang, Z.Tang and Z. Li. *Magma-dynamics of Granitoids: Theory, Method and a Case Study of the Eastern Hubei Granitoids.* Press of China University of Geosciences, Wuhan (in Chinese with English abstract) (1994).

6. C. Ma, K. Yang, Z. Tang, Y. Long, C. Ehlers and Alf Lindroos. Formation and differential rock uplift-exhumation of high-pressure metamorphic terrane in Dabie Mountains, Central China: Evidence from igneous rocks, *Earth Science-J. China Univ. Geosci.* 20, 516-520 (in Chinese with English abstract) (1995).

7. A. I. Okay, S. Xu and A. M. C. Sengor. Coesite from the Dabieshan eclogites, central China, *European J. Mineral.* 1, 595-598 (1989).

8. P. R. Renne, O. T. Tobisch and J. B. Saleeby. Thermochronologic record of pluton emplacement, deformation, and exhumation at Courtright shear zone, central Sierra Nevada, California, *Geology.* 21, 331- 334 (1993).

9. S. Sou, L. Sang, Y. Han and Z. You. *The Petrology and Tectonics in Dabie Precambrian Metamorphic Terranes, Central China.* Press of China University of Geosciences, Wuhan (in Chinese with English abstract) (1993).

10. A. L. Streckeisen. Plutonic rocks: classification and nomenclature recommended by the IUGS subcommission on the systematics of igneous rocks, *Geotimes.*18:10, 26-30 (1973).

11. X. Wang, J. G. Liou and H. K. Mao. Coesite-bearing eclogites from the Dabie Mountains in Central China, *Geology.* 17, 1085-1088 (1989).

12. S. M. Wickham. Crustal anatexis and granite petrogenesis during low-pressure regional metamorphism: the Trois Seigneurs massif, Pyrenees, France, *J. Petrol.* 28, 127-169 (1987).

13. S. T. Xu, A. I. Okay, S. Y. Ji, A. M. C. Sengor, W. Su, Y. C. Liou and L. L. Jiang. Diamond from Dabieshan eclogite and its implication for tectonic setting, *Science.* 256, 80-82 (1992).

14. D. Ye, D. Li, G. Tong and X. Qiou. On the 3T-phengite and C-type eclogite in the Xinyang metamorphic belt and their tectonic significance. In: *Formation and Development of the North China Fault Block Region* Science Press, pp. 122-132 (in Chinese). Beijing (1980).

15. R. Zhang, J. G. Liou and X. Wang. Discovery of coesite eclogites in Henan Province, Central China and its tectonic implication, *Acta Petrologica Sinica.* 9, 186-191 (in Chinese with English abstract)(1993).

16. X. Zhou and F. Wang. *Thermodynamics in Petrology.* Henan Sci. Techno. Press, Zhengzhou (in Chinese)(1987)

Proc. 30ᵗʰ Int'l. Geol. Congr., Part 15   pp. 109 – 120
Li *et al.* (Eds)
© VSP 1997

Extreme Mantle Source Heterogeneities Beneath the Northern East Pacific Rise — Trace Element Evidence From Near-Ridge Seamounts

YAOLING NIU

Department of Earth Sciences, The University of Queensland, Brisbane, Qld 4072, Australia

RODEY BATIZA

Department of Geology and Geophysics, University of Hawaii, Honolulu, HI 96822, U.S.A

Abstract

In this paper the authors summarize the results of their trace element study of lavas from 50 near-ridge seamounts on the flanks of the East Pacific Rise (EPR) between 5°N and 15°N. These seamount lavas are dominated by depleted N-type mid-ocean ridge basalts (MORB) with variably enriched E-type MORB and some extremely enriched ones resembling average compositions of ocean island basalts (OIB). This large compositional variation reflects with great fidelity the mantle source heterogeneity that is masked in lavas erupted at the EPR axis. In terms of incompatible trace element abundances, this source heterogeneity can be readily envisioned as being due to the presence of enriched domains of variable size and unevenly distributed within the ambient depleted mantle. The geochemical consequence of melting such a heterogeneous source is to produce apparent mixing relationships in the lavas. The enriched domains may be dikes or veins resulting from low-degree melt metasomatism. The low degree melts may be genetically related to eastward asthenospheric flow of Hawaii plume materials towards the EPR, as suggested by mantle tomographic studies. Trace element data suggest that the enriched materials (hence Hawaii plume materials) are ultimately derived from recycled oceanic crust.

Keywords: seamounts, East Pacific Rise, basalt, trace elements, mantle source heterogeneity

INTRODUCTION

Plate tectonics theory has established that mid-ocean ridges are mostly a passive feature in the sense that mantle upwelling beneath ridges is caused by plate separation [1-3]. Mid-ocean ridge basalts (MORB), which represent an end product of pressure-release melting of the upwelling mantle, thus record the geochemical signatures of the uppermost mantle. In comparison with basalts from elsewhere in the oceanic or continental volcanic provinces, MORB as a whole show remarkably small geochemical variations characterized by low abundances of incompatible elements, low radiogenic Sr and Pb, and high radiogenic Nd [4-8]. These observations have led to the notion that oceanic upper mantle is relatively uniformly depleted in incompatible elements; it has been designated as depleted MORB mantle (DMM) [5]. Nevertheless, the DMM is by no means compositionally uniform [9], even from ridges thermally unaffected by any known hotspots such as the East Pacific Rise (EPR) [10-19]. In fact, studies of near EPR seamounts revealed small scale yet large

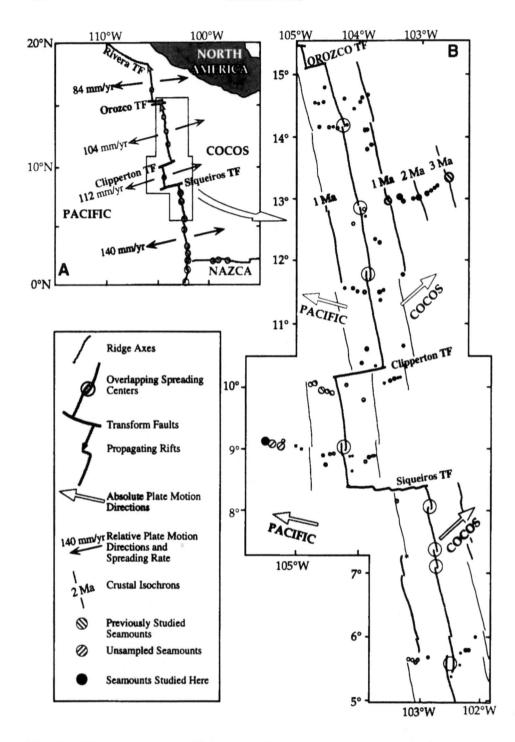

Figure 1. (a) The general tectonic framework of the northern East Pacific Rise and the vicinity. (b) A simplified map of the study area showing the locations of the near-ridge seamounts we studied.

amplitude compositional variations in the sub-EPR mantle [20-24]. In this paper we extended these studies based on our newly available trace element data on lavas from 50 near-EPR seamounts between 5°N to 15°N on both the Pacific and Cocos plates (Fig. 1). We first show that these seamount lavas span extreme compositional variations with extents of depletion and enrichment surpassing the known range of lavas from the seafloor. We then discuss the implications of the data in the context of mantle convection and ocean ridge dynamics.

RESULTS

Figure 2 shows chondrite-normalized rare-earth element (REE) abundances of the seamount lavas. The average continental crust (CC), ocean island basalts (OIB), and both enriched E-type and depleted N-type MORB are plotted for comparison. Clearly, these seamount lavas display a considerably large range of variations from extremely light-rare-earth-element (LREE) depleted samples to highly LREE-enriched basalts resembling the average OIB. Except for the few highly evolved samples (MgO < 6 wt. %; the dashed lines), the majority of the samples define a fairly simple fan-shaped pattern with more depleted samples being more depleted in the more incompatible elements and the more enriched samples having higher abundances of the more incompatible elements. The depleted samples are more depleted than samples from the Lamont Seamount chain near the 10°N (Fig. 1) [24]. No doubt, such a huge variation can only be explained by melting a mantle that is extremely heterogeneous. Figure 3 plots $[La/Sm]_{CN}$, a useful measure of the extent of source depletion or enrichment, as a function of latitude (top) and longitude (bottom) of sample locations (Fig. 1) to show that mantle source enrichment/depletion has no geographic systematics. In fact, both enriched and depleted samples can be found on the same seamounts. This indicates that the scale of the

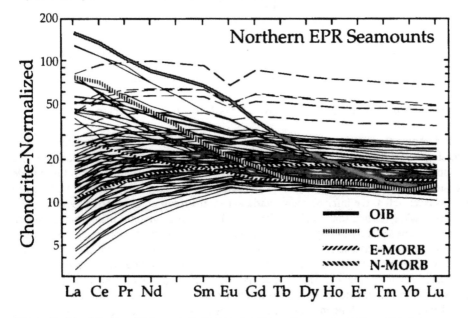

Figure 2. Chondrite-normalized rare-earth element abundances of seamount samples. Also shown are average compositions of Ocean Island Basalt (OIB), enriched E-MORB, depleted N-MORB [8], and continental crust (CC) [63] for comparison.

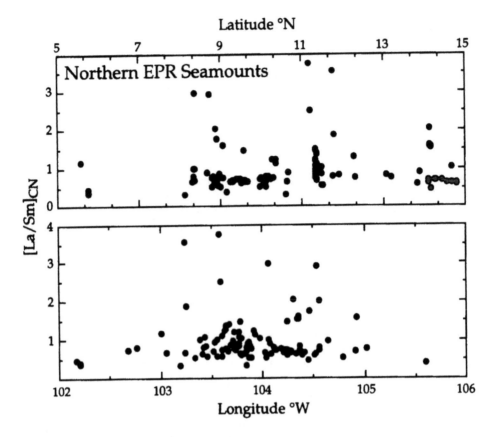

Figure 3. Chondrite-normalized [La/Sm]$_{CN}$ ratio of seamount samples are plotted as a function of latitude (top) and longitude (bottom) of seamounts studied. Clearly, the lack of systematic geographic variation, and the fact that both highly depleted and enriched lavas can be found within small areas or even single seamounts, indicates that the scales of source heterogeneities are quite small and that their distribution is spatially not uniform.

heterogeneities may be very small, perhaps on the order of 100's of meters or even smaller.

Figure 4 shows the relative variability [6] of each element for the seamount lavas. The decreasing variability is in fact consistent with decreasing relative incompatibility (or increasing bulk distribution coefficient, D) of these elements determined by the simple relationship shown in the inset [25]. This indicates that the large compositional variation in these lavas as well as the inferred source variation are the result of magmatic processes. The processes leading to such small scale yet large amplitude variation in the mantle source region may be examined closely through incompatible element ratio-ratio diagrams such as those in Figure 5. These hyperbolic curves are consistent with a binary mixing [26-27], and are also qualitatively consistent with various extents of melting given the relative incompatibility of these paired elements (Fig. 4). That is, qualitatively, samples plotting in the upper left corner represent the lowest extents of melting whereas samples in the lower right corner represent the highest extents of melting in these diagrams. It is important to note, however, that melting of a compositionally uniform source, alone, cannot explain the large amplitude variations. Further, this mixing cannot be simple binary mixing of melts because the lavas are from 50

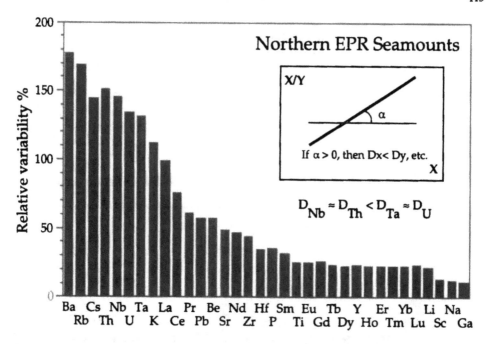

Figure 4. Relative variability of trace element abundances in seamount lavas plotted against their order of incompatibility determined by the simple relationship in the inset [25] to show the excellent correspondence between the variability and the incompatibility of elements. This indicates that the process leading to the observed trace element variations are partial melting.

different seamounts of varying ages and it is difficult to imagine the physical mixing of two singular melts over such a wide spatial and temporal interval. The curved trends in Figure 5 can be readily explained by melting a mantle that is compositionally heterogeneous via melting-induced mixing [28]. Enriched materials have lower melting temperature and thus tend to enter the melt first upon melting [21,29]. With progressive melting, the amount of the enriched material in the melt decreases as a result of dilution. Therefore, the geochemical consequence of melting of a heterogeneous mantle source is to produce the apparent mixing relationships seen in Figure 5. The important message here is that the apparently complex source heterogeneity reflected by the seamount lavas is in fact rather simple and can be explained by the presence of incompatible element enriched domains of variable size distributed widely but unevenly in the ambient depleted MORB mantle.

DISCUSSION

Why do near-ridge seamount lavas show larger geochemical variations than MORB lavas from the ridge axis?

The geochemical variations of northern EPR axial lavas have been well documented [10-19, 28, 30-32]. These variations are significantly smaller in amplitude than those seen in the nearby seamounts (Fig. 2). This difference can be explained by the action of two well-documented mixing processes that occur beneath the ridge axis but not under seamounts. First, MORB from the ridge axis represents melting of a large volume in the mantle. Melt

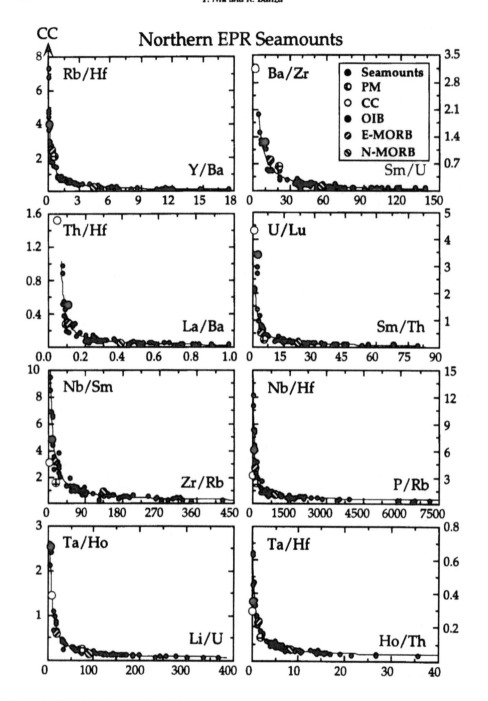

Figure 5. Plots of various highly and moderately incompatible element pairs to show that the seamount data can be explained by melting a source that has two reasonably uniform components: an enriched component and a depleted endmember. The hyperbola are due to melting-induced mixing, not mixing of two melts. For comparison, average continental crust (CC) [63], primitive mantle (PM), OIB, and both E- and N-MORB [8] are also plotted.

migration and focussing towards the very narrow (1-2 km) axial accretion zone [33] are an important mixing process that homoginizes melts moving upwards and laterally towards the axis. Secondly, additional mixing occurs in axial magma chambers that exist along much of the EPR axis [33-35] further homoginizing the melt prior to eruption [20-21, 23, 28, 30, 32]. In contrast, off-axis seamounts represent much smaller melt volumes tapped locally [20-21, 23, 28, 30, 32, 36-37], and lack steady-state magma chambers. Because they are much less efficiently mixed, seamount lavas reflect with greater fidelity than axial lavas the actual mantle source heterogeneity beneath the EPR.

Characteristic of the enriched heterogeneity

The above observations place three limits on the enriched heterogeneities: (1) the observed mixing is a melting-induced mixing, i.e., the enriched heterogeneities exist as physically distinct domains in the ambient depleted mantle prior to the major melting events (Fig. 5); (2) the sizes of the enriched heterogeneities must be variably small, and their distribution is not uniform (Fig. 3); and (3) the enriched heterogeneities must be of low-degree melt origin because the enriched samples have higher abundances of more incompatible elements (Fig. 2) and the relative abundance variability of trace elements is proportional to their relative incompatibility (Fig. 4). These observations, taken together, suggest that the enriched heterogeneities exist in the immediate source region in the form of small dikes or veins [28, 38-39]. The small sizes of the enriched heterogeneities are required to explain the coexistence of both depleted and enriched lavas within geographically small areas (Fig. 3) such as single seamounts [20-24, 28, 36-37]. The question yet to be answered is the origin of the low-degree melts that we suggest to occur in the form of dikes or veins. From trace elements alone (Figs. 2 and 5), it is obvious that the enriched component is broadly similar to the source for OIB, but the apparent absence of any known hotspots in the northern EPR region requires another mantle process, in addition to simple passive upwelling, to effectively transport plume materials to the sub-ridge mantle beneath the northern EPR.

Where does the plume material come from and why are the enriched heterogeneities beneath the EPR of wide but spotty dispersal?

The presence of enriched plume-like material beneath the northern EPR region has been puzzling because there are no known hotspots in this broad EPR region. However, recent mantle tomographic studies [40-41] show that lateral asthenospheric flow of plume materials towards ocean ridges is likely to be a wide-spread phenomenon. Figure 6 shows that indeed there is an obvious low-velocity layer at ~ 100 to 250 km depth beneath Hawaii that extends laterally towards the northern EPR. While the large distance (~ 5500 km) between Hawaii and the northern EPR makes such a link seem doubtful at first glance, large scale lateral flow of asthenosphere (even counter-flow) is apparently required to explain some geoid anomalies [41-42]. If the Hawaiian plume is, in fact, the source of the enriched component in the area of the northern EPR, the great distance of transport helps to explain both the spotty dispersal of the enriched heterogeneities and the absence of any thermal effects of the plume, as seen at ridges located nearer active plumes [43-47]. While large scale melting of plume material may not take place in the course of lateral asthenospheric flow because of little or limited decompression, very low-degree melts of low-melting point components must inevitably form and metasomatize the ambient depleted mantle. We propose that it may be this process that explains the enriched component of the seamount lavas.

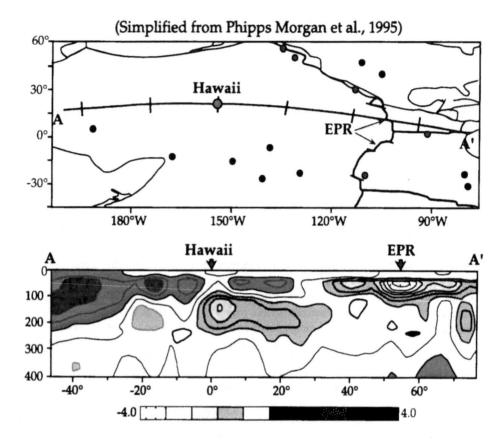

Figure 6. The top panel shows a portion of the global hotspot distribution (solid circles) and below is a vertical cross section of seismic shear velocity anomalies along a great circle path that connects the Hawaii hotspot and the EPR (profile A-A'). Note that there exists a clear low velocity layer at ~ 100 - 250 km depth beneath Hawaii that extends towards the East Pacific Rise to the east. Total velocity variation is ± 4%. Contour lines are at half the shading interval. Taken from [41].

The ultimate source of the enriched component or Hawaiian plume material

It is generally agreed that mantle convection and crustal recycling are the primary mechanisms for creating enriched heterogeneities [48-52] in the deep mantle that rise as plume sources to supply OIB. However the exact location of these OIB reservoirs remains open to debate [5, 25, 53-57]. A curious question is whether our data have sufficient resolution to decipher the ultimate source of the enriched component beneath the northern EPR. A potential clue is provided by the fact that, for seamounts, we find $D_{Nb} \approx D_{Th} < D_{Ta} \approx D_U$ (Fig. 4). That is, mantle melting beneath seamounts (and also the EPR) does not fractionate Nb from Th, nor Ta from U; any Ta and Nb anomalies in lavas must, therefore, be a source signature inherited from previous events. Figure 7 shows that the mantle sources for the seamount lavas possess excess Ta and Nb. Despite the scatter, the data define a significant trend with generally more enriched lavas having higher excess Ta and Nb than depleted lavas. Several important implications of Figure 7 are (1) the missing Nb and Ta in the continental crust clearly must reside in the mantle source for oceanic basalts, and there is

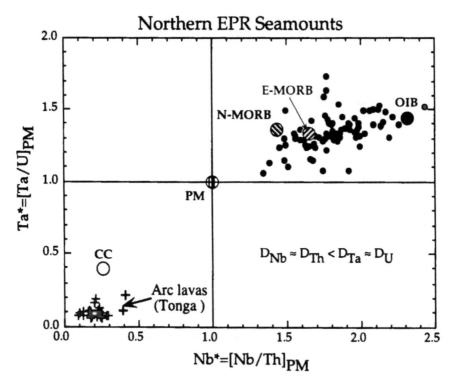

Figure 7. Ta* versus Nb* for the near-EPR seamount lavas. Average continental crust (CC) [63], primitive mantle (PM), OIB, and both E- and N-MORB [8], and unpublished Tonga arc lava data of A. Ewart are also plotted for comparison. Given the $D_{Nb} \approx D_{Th} < D_{Ta} \approx D_U$ relationship during mantle melting beneath the ridge, the excess Ta and Nb in oceanic basalts are inherited from their sources.

no need to invoke a hidden Nb-Ta rich reservoir deep in the mantle [7,58]; (2) recycled continental crust material is unlikely to be significant in the source region of the oceanic basalts, being too depleted in Nb and Ta; (3) as subduction zone related arc magma genesis is the only known process that fractionates Nb from Th, and Ta from U (e.g., see Tonga arc lavas), it appears clear that subduction-zone processes must also be responsible for the excess Ta and Nb in the source region of oceanic basalts; and (4) recycled oceanic crust is, therefore, most likely the ultimate source of Hawaiian plume material (e.g., the Koolau volcanics, which also possess excess Nb relative to Th [59-60]) as proposed previously [6-7, 61], and hence the enriched component beneath the northern EPR. Nb and Ta may not partition into aqueous fluid removed from the down-going slab as effectively as the low field strength elements, resulting in their relative enrichments in subducted materials [7-8, 58, 62].

CONCLUSIONS

Near-EPR seamount lavas are compositionally quite variable, reflecting with great fidelity of mantle source heterogeneity that is not so obvious in the nearby EPR axial lavas. This heterogeneity is characterized by the presence of incompatible element enriched domains of variably small size non-uniformly distributed within the ambient depleted mantle. The enriched domains most likely exist as dikes or veins resulting from low-degree melt "metasomatism". These low-degree melts may have been derived from Hawaiian plume materials that flow at the asthenospheric level towards the northern EPR. The remote distance between Hawaii and the EPR explains both normal (thermally unaffected) EPR topography and the wide but spotty dispersal of the enriched heterogeneities present beneath the EPR. The observation that $D_{Nb} \approx D_{Th} < D_{Ta} \approx D_U$ and the excess Nb-Ta relative to Th-U in the seamount lavas suggest that the ultimate source of the enriched material (i.e., the source of Hawaii plume) is likely to be recycled oceanic crust.

Acknowledgments

Supports by the US NSF and ONR, the Australian ARC, and The University of Queensland are gratefully acknowledged. We thank Tony Ewart for unpublished Tonga trace element data used in Figure 7.

REFERENCES

1. D. McKenzie. Some remarks on heat flow and gravity anomalies, *Jour. Geophys. Res.* 72, 6261-6273 (1967).
2. B. Parsons and J.G. Sclater. An analysis of the variation of ocean floor bathymetry and heat flow with age, *Jour. Geophys. Res.* 82, 803-827 (1977).
3. D. McKenzie and M.J. Bickle. The volume and composition of melt generated by extension of the lithosphere, *Jour. Petrol.* 29, 625-679 (1988).
4. W.M. White. Sources of oceanic basalts: Radiogenic isotope evidence, *Geology* 13, 115-118 (1985).
5. A. Zindler and S.R. Hart. Chemical geodynamics, *Ann. Rev. Earth Planet. Sci.* 14, 493-571 (1986).
6. A.W. Hofmann. Chemical differentiation of the Earth: the relationship between mantle, continental crust, and oceanic crust, *Earth Planet. Sci. Lett.* 90, 297-314 (1988).
7. A.D. Saunders, M.J. Norry and J. Tarney. Origin of MORB and chemically-depleted mantle reservoirs: trace element constraints, *Jour. Petrol.* 29, 415-445 (1988).
8. S.-S. Sun and W.F. McDonough. Chemical and isotopic systematics of ocean basalt: Implications for mantle composition and processes. In: *Magmatism of the ocean basins.* A.D. Saunders and M.J. Norry (Eds). pp. 323-345. Soc. London Spec. Publ. 42, London (1989).
9. W.G. Melson, T.L. Vallier, T.L. Wright, G. Byerly and J. Nelen. Chemical diversity of abyssal volcanic glass erupted along Pacific, Atlantic and Indian Ocean floor spreading centers, *Amer. Geophys. Un. Mon.* 19, 351-368 (1976).
10. R. Batiza, B.R. Rosendahl and R.L. Fisher. Evolution of oceanic crust, 3, Petrology and chemistry of basalts from the East Pacific Rise and Siqueiros transform fault, *Jour. Geophys. Res.* 82, 265-276 (1977).
11. C.H. Langmuir, J.F. Bender and R. Batiza. Petrological and tectonic segmentation of the East Pacific Rise, 5°30'- 14°30'N, *Nature* 332, 422-429 (1986).
12. W.M. White, A.W. Hofmann and H. Puchelt. Isotope geochemistry of Pacific mid-ocean ridge basalt, *Jour. Geophys. Res.* 92, 4881-4893 (1987).
13. R. Hékinian, G. Thompson and D. Bideau. Axial and off-axial heterogeneity of basaltic rocks from the East Pacific Rise at 12°35'N-12°51'N and 11°26'N-11°30'N, *Jour. Geophys. Res.* 94, 17,437-17,463 (1989).
14. A. Prinzhofer, E. Lewin and C.J. Allègre. Stochastic melting of the marble cake mantle: evidence from local study of the East Pacific Rise at 12°50, *Earth Planet. Sci. Lett.* 92, 189-106 (1989).
15. J.H. Natland. Partial melting of a lithologically heterogeneous mantle: Inferences from crystallisation histories of magnesian abyssal tholeiites from the Siqueiros Fracture Zone. In: *Magmatism of the ocean basins* A.D. Saunders and M.J. Norry (Eds). pp. 41-70. Soc. London Spec. Publ. 42, London (1989).
16. J.M. Sinton, S.M. Smaglik, J.J. Mahoney and K.C. Macdonald. Magmatic processes at superfast spreading mid-ocean ridges: glass compositional variations along the East Pacific Rise 13°-23°S, *Jour. Geophys. Res.* 96, 6133-6155 (1991).
17. J.R. Reynolds, C.H. Langmuir, J.F. Bender, K.A. Kastens and W.B. F. Ryan. Spatial and temporal variability in the geochemistry of basalts from the East Pacific Rise, *Nature* 359, 493-499 (1992).
18. J.J. Mahoney, J.M. Sinton, D.M. Kurz, J.D. Macdougall, K.J. Spencer and G.W. Lugmair. Isotope and trace

element characteristics of a super-fast spreading ridge: East Pacific Rise, 13 - 23°S, *Earth Planet. Sci. Lett.* **121**, 173-193 (1994).

19. W. Bach, E. Hegner, E. Erzinger and M. Satir. Chemical and isotopic variations along the superfast spreading East Pacific Rise from 6° to 30°S, *Contrib. Mineral. Petrol.* **116**, 365-380 (1994).

20. R. Batiza and D.A. Vanko. Petrology of young Pacific seamounts, *Jour. Geophys. Res.* **89**, 11,235–11,260 (1984).

21. A. Zindler, H. Staudigel and R. Batiza. Isotope and trace element geochemistry of young Pacific seamounts: Implications for the scale of upper mantle heterogeneity, *Earth Planet. Sci. Lett.* **70**, 175-195 (1984).

22. D.W. Graham, A. Zindler, M.D. Kurz, W.J. Jenkins, R. Batiza and H. Staudigel. He, Pb, Sr, and Nd isotope constraints on magma genesis and mantle heterogeneity beneath young Pacific seamounts, *Contrib. Mineral. Petrol.* **99**, 446-463 (1988).

23. R. Batiza, Y. Niu and W.C. Zayac. Chemistry of seamounts near the East-Pacific Rise: Implications for the geometry of sub-axial mantle flow, *Geology* **18**, 1122-1125 (1990).

24. D.J. Fornari, M.R. Perfit, J.F. Allan, R. Batiza, R. Haymon, A. Barone, W.B.F. Ryan, T. Smith, T. Simkin and M. Luckman. Geochemical and structural studies of the Lamont seamounts: Seamounts as indicators of mantle processes, *Earth Planet. Sci. Lett.* **89**, 63–83 (1988).

25. A.W. Hofmann, K.P. Jochum, M. Seufert, and W.M. White. Nb and Pb in oceanic basalts: new constraints on mantle evolution, *Earth Planet. Sci. Lett.* **79**, 33-45 (1986).

26. R. Vollmer. Rb-Sr and U-Th-Pb systematics of alkaline rocks: the alkaline rocks from Italy, *Geochim. Cosmochim. Acta* **40**, 283-295 (1976).

27. C.H. Langmuir, R.D. Vocke and G.N. Hanson. A general mixing equation with application to Icelandic basalts, *Earth Planet. Sci. Lett.* **37**, 380-392 (1978).

28. Y. Niu, D.G. Waggner, J.M. Sinton and J.J. Mahoney. Mantle source heterogeneity and melting processes beneath seafloor spreading centers: the East Pacific Rise, 18°-19°S, *Jour. Geophys. Res.* **101** (in press).

29. N. H. Sleep. Tapping of magmas from ubiquitous heterogeneities: An alternative to mantle plumes?, *Jour. Geophys. Res.* **89**, 10,029-10,041 (1984).

30. R. Batiza and Y. Niu. Petrology and magma chamber processes at the East Pacific Rise ~ 9°30'N, *Jour. Geophys. Res.* **97**, 6779-6797 (1992).

31. M.R. Perfit, D.J. Fornari, M.C. Smith, J.F. Bender, C.H. Langmuir and R.M. Haymon. Small-scale spatial and temporal variations in mid-ocean ridge crest magmatic processes, *Geology* **22**, 375-379 (1994).

32. R. Batiza, Y. Niu, J.L. Karsten, W. Boger, E. Potts, L. Norby and R. Butler. Steady and non-steady state magma chambers below the East Pacific Rise, *Geophys. Res. Lett.* **23**, 221-224 (1996).

33. K.C. Macdonald, P.J. Fox, L.J. Perram, M.F. Eisen, R.M. Haymon, S.P. Miller, S.M. Carbotte, M.-H. Cormier, and A.N. Shor. A new view of the mid-ocean ridge from the behavior of ridge-axis discontinuities, *Nature* **335**, 217-225 (1988).

34. R.S. Detrick, J.P. Madsen, P.E. Buhl, J, Vera, J. Mutter, J. Orcutt and T. Brocker. Multichannel seismic imaging of an axial magma chamber along the East Pacific Rise between 4°N and 13°N, *Nature* **326**, 35–41 (1987).

35. J.M. Sinton and R. S. Detrick. Mid-ocean ridge magma chambers: *Jour. Geophys. Res.* **97**, 197-216 (1992).

36. R. Batiza, T.L. Smith and Y. Niu. Geologic and Petrologic evolution of seamounts near the EPR based on submersible and camera study, *Mar. Geophys. Res.* **11**, 169–236 (1989).

37. Y. Niu and R. Batiza. An empirical method for calculating melt composi-tions produced beneath mid-ocean ridges: application for axis and off-axis (seamounts) melting. *Jour. Geophys. Res.* **96**, 21,753-21,777 (1991).

38. D.A. Wood. A variably veined suboceanic upper mantle - Genetic significance for mid-ocean ridge basalts from geochemical evidence, *Geology* **7**, 499-503 (1979).

39. A.P. Le Roex, H.J.B. Dick, A.L. Erlank, A.M. Reid, F.A. Frey and S.R. Hart. Geochemistry, mineralogy and petrogenesis of lavas erupted along the Southwest Indian Ridge between the Bouvet Triple Junction and 11 degrees east, *Jour. Petrol.* **24**, 267–318 (1983).

40. Y.-S. Zhang and T. Tanimoto. high resolution global upper mantle structure and plate tectonics, *Jour. Geophys. Res.* **98**, 9793-9823 (1993).

41. J. Phipps Morgan, W.J. Morgan, Y.-S. Zhang and W.H.F. Smith. Observational hints for a plume-fed, suboceanic asthenosphere and its role in mantle convection, *Jour. Geophys. Res.* **100**, 12,753 - 12,767 (1995).

42. J. Phipps Morgan and W.H.F. Smith. Flattening of the seafloor depth-age curves as a response to asthenospheric flow, *Nature* **359**, 524-527 (1992).

43. S.-S. Sun, M. Tatsumoto, and J.-G. Schilling. Mantle plume mixing along the Reykjanes ridge axis: Lead isotope evidence, *Science* **190**, 143-147 (1975).

44. J.-G. Schilling, M. Zajac, R. Evans, T. Johnston, W. White, J.D. Devine and R. Kingsley. Petrological and geochemical variations along the Mid-Atlantic Ridge from 29°N to 73°N, *Amer. Jour. Sci.* **283**, 510–586 (1983).

45. J.-G. Schilling. Fluxes and excess temperatures of mantle plumes inferred from their interaction with migrating mid-ocean ridges, *Nature* **352**, 397-403 (1991).

46. H. Bougault, L. Dmitriev, J.-G. Schilling, A. Sobolev, J.L. Jordan and H.D. Needham. Mantle heterogeneity from trace elements: MAR triple junction near 14°N, *Earth Planet. Sci. Lett.* **88**, 27–36 (1988).

47. L. Dosso, H. Bougault and J.-L. Joron. Geochemical morphology of the North Mid-Atlantic Ridges, 10°-24°N: Trace element-isotope complementary, *Earth Planet. Sci. Lett.* **120**, 443-462 (1993).

48. G.N. Hanson. Geochemical evolution of the suboceanic mantle, *Jour. Geol. Soc. London* **134**, 235-253 (1977).

49. C.J. Allègre, B. Hamelin and B. Dupré. Statistical analyses of isotopic ratios in MORB: the mantle blob cluster model and the convective regime of the mantle, *Earth Planet. Sci. Lett.* **71**, 71-84 (1984).

50. J.G. Fitton and D. James. Basic volcanism associated with intraplate linear features, *Philos. Trans. R. Soc. London Ser.* **A317**, 253-266 (1986).

51. J.J. Mahoney, J.H. Natland, W.M. White, R. Poreda, S.H. Bloomer, R.L. Fisher and A.N. Baxter. Isotopic and geochemical provinces of the western Indian Ocean spreading centres, *Jour. Geophys. Res.* **94**, 4033–4052 (1989).

52. B.L. Weaver. The origin of ocean island basalt end-member compositions: trace element and isotopic constraints, *Earth Planet. Sci. Lett.* **104**, 381-397 (1991).

53. W.M. White, and A.W. Hofmann. Sr and Nd isotope geochemistry of oceanic basalts and mantle evolution, *Nature* **296**, 821-825 (1982).

54. B. Dupre and C.J. Allègre. Pb-Sr isotope variation in Indian Ocean basalts and mixing phenomena, *Nature* **303**, 142-146 (1983).

55. R.N. Thompson, G.L. Hendry and S.J. Parry. An assessment of the relative roles of crust and mantle in magma genesis: An elemental approach, *Philo. Trans. R. Soc. London* **A310**, 549-590 (1984).

56. A.E. Ringwood. Mantle dynamics and basalt genesis, *Tectonophysics* **112**, 17-34 (1985).

57. S.R. Hart and A. Zindler. Constraints on the nature and development of chemical heterogeneities in the mantle. In: *Mantle Convection.* W.R. Peltier (Ed). pp. 262-387. Gordon and Breach Science Publishers (1989).

58. W.F. McDonough. Partial melting of subducted oceanic crust and isolation of its residual eclogitic lithology, *Phil. Trans. R. Soc. London* **A 335**, 407-418 (1991).

59. F.A. Frey, M.O. Garcia and M.F. Roden. Geochemical characteristics of Koolau Volcano: Implications of inter shield geochemical difference among Hawaiian volcanoes, *Geochim. Cosmochim. Acta* **58**, 1441-1462 (1994).

60. M.F. Roden, T. Trull, S.R. Hart and F.A. Frey. New He, Nd, Pb, and Sr isotopic constraints on the constitution of the Hawaiian plume: Results from Koolau Volcano, Oahu, Hawaii, USA, *Geochim. Cosmochim. Acta* **58**, 1431-1440 (1994).

61. A.W. Hofmann and W.M. White. Mantle plumes from ancient oceanic crust, *Earth Planet. Sci. Lett.* **57**, 421-436 (1982).

62. J. A. Pearce and D.W. Peate. Tectonic implications of the composition of volcanic arc magmas, *Ann. Rev. Earth Planet. Sci.* **23**, 251-285 (1995).

63. R.L. Rudnick and D.M. Fountain. Nature and composition of the continental crust: a lower crustal perspective, *Rev. Geophys.* **33**, 267-309 (1995).

Proc. 30ᵗʰ Int'l Geol. Congr., Part , pp. 121 – 135
Li et al.. Eds.)

Processes in a Composite, Recharging Magma Chamber: Evidence from Magmatic Structures in the Aztec Wash Pluton, Nevada

DAVID W. PATRICK* and CALVIN F. MILLER

Department of Geology, Vanderbilt University, Nashville, TN 37235 U.S.A.
*Presently at:
Department of Science, Brockton High School , Brockton , MA 02401 U.S.A.

Abstract

Magmatic structures in the Aztec Wash pluton preserve a record of repeated injections of mafic magma into a felsic magma chamber and provide information about subsequent tilting. The pluton is divided into a southern (lower) heterogeneous zone (HZ), which contains the abundant magmatic structures, and a northern (upper) homogeneous granite (HG), which is virtually structureless. HG is entirely felsic (most ~72 wt.% SiO_2), whereas the HZ is extremely diverse in composition (44-76 wt% SiO_2).

Sequences of meter-scale mafic sheets with intervening felsic material mark repeated injection of mafic magma into the HZ. The sheets spread laterally and ponded, trapping more felsic magma or crystal mush beneath them. Most of the sheets are fine-grained with delicately crenulated, quenched margins against felsic rock, but thick sheets apparently cooled slowly enough to crystallize coarse cumulates. Gravitationally unstable felsic melt in the trapped felsic layers intruded the overlying mafic layers to form vertical sheets and pipes. The lateral spreading of the mafic sheets was interrupted near the margins of the HZ, where magma injected and disaggregated the wall rock to form abundant xenolithic blocks. After the HZ was semisolid, composite vertical sheets with mafic pillows in a felsic matrix intruded, apparently as a coarse slurry, in response to fresh injections of mafic magma into the base of the chamber. A gradational contact with the overlying HG suggests that HG was still at least partly molten when late HZ magmas intruded, but preservation of significant topography on the HZ-HG surface and of trains of mafic material above the contact indicate that it had appreciable strength.

The structures in the HZ demonstrate the complex, protracted way in which magma chambers can fill. This provides a mechanism for maintaining a chamber in a partially molten state for a much longer interval than if it had been emplaced all at once, or had contained only mafic or felsic magma. Furthermore, because many of the structures are inferred to have been either horizontal or vertical when they formed, they provide a means of estimating crustal tilting and therefore of reconstructing the original geometry of the pluton. The tilt estimates also clarify important extensional structures in this area that were difficult to define because of the absence of stratified rocks.

Keywords: composite pluton, magma chamber, granite, magmatic structures, recharging, tilt indicators

INTRODUCTION

Background and Purpose

A fundamental goal of petrology is to better understand what active magma chambers in continental crust are like and how they work. Numerous questions about magma chambers have have provoked considerable controversy in recent years - for example, what is the volume of magma present at any time, and how much does it fluctuate?; how influential is mafic injection into felsic magma chambers?; what is the longevity of magma chambers, and do they commonly wax and wane thermally and volumetrically?; are they effectively stirred by chamber-wide convection? Some of the best constraints on these questions, at least for individual examples, can come from field-based studies of heterogeneous plutons that record interaction between contrasting magmas [e.g. 1,3,5,6,12,17,18,27,28]. Mafic-felsic magma interaction has the potential to dramatically alter the nature and history of magma chambers.

Extensive three-dimensional exposure of mafic, intermediate, and felsic rock in the Aztec Wash pluton of southern Nevada provides an outstanding opportunity to describe and interpret structures that mark interactions between contrasting magmas. The protracted, complex chamber history that can be demonstrated here bears upon questions of how magma chambers can operate and illustrates the importance of understanding magmatic structures.

Geologic Setting

The Eldorado Mountains are located in the northwestern corner of the Colorado River extensional corridor (CREC [15]). The CREC is a northward trending 50 to 100 km wide zone of moderately to severely extended crust in the Basin and Range province south of Las Vegas, Nevada (Fig. 1a). Extension here began ~16-18 Ma and terminated sometime between 10 and 14 Ma [e.g. 11,13,14]. Intrusions make up a majority of the crust that underlies roughly coeval volcanic strata in the Eldorado Mountains [19]. The 15.7 Ma Aztec Wash pluton, which is the youngest of the plutons in the range, intrudes volcanic strata, Proterozoic basement, and slightly older intrusions (Fig. 1b)[8]. The pluton straddles a major accommodation zone that separates east-tilted structural blocks to the north from west-tilted blocks to the south [e.g. 9-11]. Geometry and tilting of the pluton are thus important to interpretation of the kinematics of extension in this region.

Falkner and coworkers [7,8] have described the petrology of the Aztec Wash pluton. They noted that the pluton is divided into a northern portion comprised of homogeneous granite, and a southern portion that is highly heterogeneous, including mafic, felsic, and intermediate rocks, where there is widespread evidence of magma mingling. They concluded that the exposed rocks represent at least two discrete magmas, one derived from enriched mantle and the other rich in ancient crustal component. Multiple pulses of each magma type intruded to form the composite magma chamber. Falkner noted the evidence for magma mingling but worked mostly in the northern portion of the heterogeneous zone and in the granite, and they did not emphasize the diverse magmatic structures upon which we focus in this paper.

GEOLOGY OF THE HETEROGENEOUS ZONE
Introduction

Falkner [7,8] recognized the diversity and clear evidence for open system processes and multiple injection in the southern area of the pluton that she referred to as the Heterogeneous Zone (HZ), but she did not distinguish discrete generations of injection and mingling of contrasting magmas. She also did not observe some of the more spectacular and meaningful structures that are present in the southern part of the HZ. She did note that the rocks of the HZ and the Homogeneous Granite (HG) to the north appear to be broadly coeval, and that mafic and felsic dikes postdate both (though they may predate final solidification). We have studied the mingled magmas represented by the HZ in more detail and identified two major units: earlier, more voluminous intrusions with gently to moderately dipping primary fabric (Unit 1), and later, cross-cutting intrusions with subvertical fabric (Unit 2)[20-22]. Both include mafic and felsic rocks and diverse rocks of intermediate composition. In this study we emphasize description and interpretation of the magmatic structures within these units and how they bear upon the history of the Aztec Wash magma chamber.

Following the work of Falkner [7,8] and Patrick [20], we divide the Aztec Wash pluton into the following units:
Homogeneous Granite (HG), which constitutes all of the northern 40% of the pluton. Except for small very felsic dikes and a discontinuous felsic border zone, this rock is uniformly a medium-grained, structureless K-feldspar-rich granite with subordinate subhedral plagioclase, interstitial quartz, minor biotite and prominent sphene, and about 72 wt% SiO_2 [7]. Similar granites occur within the HZ associated with Units 1 and especially 2, but we include those granites as parts of those units because they are clearly coeval and mingled with the mafic rocks of the heterogeneous zone.
Units 1 and 2, each of which include both mafic (U1m, U2m) and felsic (U1f, U2m) rocks. Both are described in detail below. They are distinguished, as noted above, primarily on the basis of orientation and cross-cutting relations. Only Unit 1 is of mappable scale; it constitutes most of the HZ. Unit 2, though widespread in the HZ, is exposed over small areas.

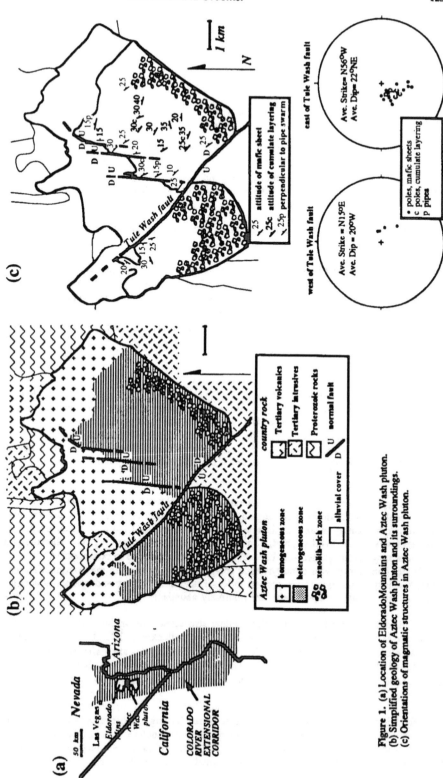

Figure 1. (a) Location of Eldorado Mountains and Aztec Wash pluton.
(b) Simplified geology of Aztec Wash pluton and its surroundings.
(c) Orientations of magmatic structures in Aztec Wash pluton.

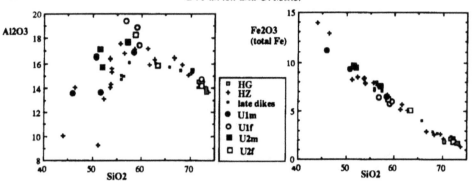

Figure 2. Variation of major element oxides vs. SiO₂ within Aztec Wash pluton. Data for HZ (Heterogeneous Zone, undivided), HG (Homogeneous Granite), and late dikes from [14]; remainder from [18].

Late dikes - north-striking mafic and felsic dikes that cross-cut the other rocks of the HZ and HG. These dikes are described briefly below.

Geochemically, the rocks of the HZ form a continuum from extremely mafic (44 wt% SiO₂, 22 wt% MgO) to the very felsic composition typical of HG granite (Figure 2). Although there are no significant compositional gaps, rocks with less than about 52 wt% SiO₂ and between 58 and 68 wt% SiO₂ are less common than rocks with 52-58 and >68 wt% SiO₂. All of the units fit within the same general compositional trend, which is highly inflected for some elements.

Units of the Heterogeneous Zone
Unit 1

Unit 1, characterized by relatively gentle dips of magmatic structures and by the fact that it predates the other units that contain mafic rocks, is the dominant rock unit of the HZ. We divide U1 into mafic (U1m) and felsic (U1f) subunits on the basis of field appearance and inferred origin; both subunits include rocks that would be geochemically characterized as intermediate. U1f rocks are lighter in color and coarser grained than U1m rocks, and they serve as host to masses of U1m. U1m rocks exceed U1f in exposed area by roughly a 3:2 or 2:1 margin in the HZ.

U1m rocks typically are fine-grained (generally phaneritic) and occur as tabular sheets with frequent reentrants (cf. boudinage), ranging in thickness from 5 cm to 2 meters and varying greatly in lateral extent; as similar discrete, flattened pillows; as 1 cm to 5 meter, irregularly shaped enclaves; and as well-aligned, 5 to 20 cm ellipsoidal enclaves (Fig. 3). The fine-grained mafic rocks range from gabbro (or diabase) to quartz monzodiorite in composition. Hornblende, commonly with clinopyroxene cores, and plagioclase are the dominant minerals. Biotite, K-feldspar, and quartz are present in lesser amounts. Sphene and apatite, typically interstitial and acicular, respectively, are relatively abundant.

Some U1m rocks are much coarser grained, either structureless or rhythmically layered. These rocks occur as relatively extensive exposures (≥10 m in thickness, 10's of m in lateral extent) or angular blocks in a U1f matrix (Fig. 4). The coarse mafic rocks are mafic gabbros that are rich in euhedral clinopyroxene ± olivine. Those without olivine contain abundant euhedral clinopyroxene, coarse, poikilitic hornblende, and plagioclase. Apatite is present as stubby prisms.

Relatively felsic, coarse-grained pockets characterized by large, euhedral hornblende crystals ("appinitic" texture) set in a matrix that is dominantly plagioclase, though volumetrically very minor, are widespread within U1m masses. Small, aplitic dikes that are probably associated with U1f cut U1m. These dikelets commonly contain miarolitic cavities.

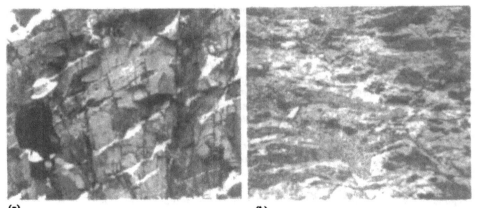

(a)   (b)

Figure 3. (a) Fine-grainedU1m (mafic) sheets with intervening trapped U1f (felsic rock). (b) Small (10 - 50 x 5-10 cm), parallel-aligned mafic enclaves.

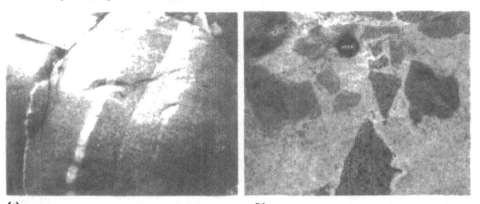

(a)   (b)

Figure 4. (a) Rhythmically layered, coarse U1m intruded by aplitic U1f. (b) Angular blocks of coarse U1m with fine-grained U1m blocks and pillows.

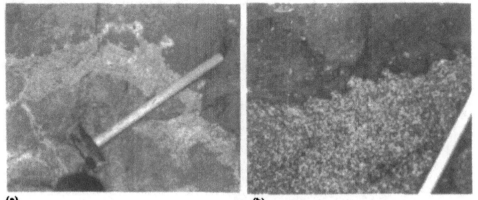

(a)   (b)

Figure 5. (a) Fine-grained U1m with crenulate margins against trapped U1f. (b) Coarse feldspars in quenched U1m. Feldspars are similar to those in adjacent U1f. (c) U1m pillows with partly angular and partly crenulate contacts against U1f.

U1f almost invariably is either host to abundant small enclaves of U1m or constitutes the matrix between larger sheets and masses of U1m (Fig. 3, 5). It does not form extensive, homogeneous exposures. U1f rocks range from quartz monzodiorite to granite, all with the assemblage plagioclase (subhedral, strongly zoned) + biotite + K-feldspar (interstitial to subhedral)+ quartz (interstitial) + accessories (abundant euhedral sphene; acicular apatite; zircon; opaques; allanite). Hornblende is common in more mafic samples, and absent in some of the most felsic; clinopyroxene is present in the more mafic samples, mostly in cores of mafic clots. U1f quartz monzodiorites differ from those of U1m in being coarser and clearly distinct from adjacent U1m rocks. They commonly have a primary magmatic fabric defined by plagioclase laths and mafic minerals that is parallel to adjacent mafic sheets. Only rarely are U1f samples as felsic as Homogeneous Granite.

The pillow-like masses and sheets that are the commonest U1m rocks have vaguely to sharply defined crenulate contacts against U1f and typically are finer-gained near contacts and coarsen inward (Fig. 5a). Some of these U1m rocks contain coarse feldspars that are identical to those in adjacent felsic rocks (Fig. 5b). The contacts of some pillow-like enclaves are in part angular; on angular sides, these pillows show no core to rim gradation in grain size. Coarse-grained enclaves are angular on all sides and lack fine-grained margins (cf. Fig. 4).

Sheet-like protrusions of U1f into fine-grained U1m sheets are common, and similar cylindrical, pipe-like protrusions are present in several areas (Fig. 6). Sheets either narrow upward and terminate in the overlying U1m sheet, or connect with an overlying U1f sheet. The pipes, which locally are very abundant, are uniformly about 10-20 cm in diameter and roughly 1 m long, flaring out where they connect with underlying U1f sheets. Pipes are uniformly perpendicular to the mafic sheets that they intrude. The pipes and sheet-like protrusions are very felsic in all cases, but they may entrain mafic material, presumably derived from surfaces of nearby pillows (Fig. 6b).

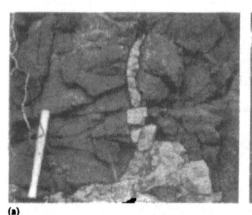

(a)

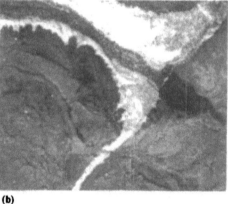

(b)

(c)

Figure 6. (a) Vertical U1f sheet intruding fine-grained U1m. (b) Vertical U1f sheet with entrained mafic material. (c) Felsic pipes intruding U1m, showing a single longitudinal and several transverse sections.

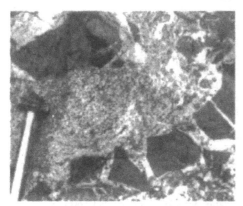

Fig. 7

Fig. 8

Figure 7. Xenolith of gneiss, cut by fine-grained mafic dikes which were fractured after emplacement. Felsic melt, probably derived from both gneiss and U1f, intruded both dikes and gneiss.

Figure 8. Composite, nearly vertical U2 dike cutting gently dipping U1m and U1f sheets.

Figure 9. Distant view of gently north-dipping Heterogeneous Zone-Homogeneous Granite contact.

Fig. 9

A zone more than one km across along the southern and southeastern margin of the pluton contains abundant gneissic xenoliths derived from the surrounding country rock. These blocks range from cm to 10's of m across, and they locally constitute as much as 40% of large exposures. Most are angular, but smaller xenoliths show evidence of incipient melting with attendant rounding and deformation. In some cases, a sequence of diking of xenolithic blocks by mafic rock, subsequent partial melting of the block, and finally blending of the anatectic melt with U1f granite can be discerned (Fig. 7). Within this xenolith-rich zone, typical rock types of U1 are present, but well-defined mafic sheets and other orderly structures are absent.

Orientation of Magmatic Structures in Unit 1- In many areas, U1m sheets have relatively uniform orientations (e.g. Fig. 3, 4). We have measured attitudes of sequences of sheets from 21 areas (Fig. 1c). East of the Tule Wash Fault, which transects the pluton from SE to NW, the fabrics display an average strike of ~N55°W and average dip of ~22° to the NE. West of the fault the sheets display an average strike of ~N15°W and dip of ~20° NW. Orientations of abundant pipes, measured in two of the areas east of the fault, are perpendicular to the layering (see Fig. 1c). Layering in four rhythmically layered gabbro sequences, also east of the fault, is essentially parallel to sheets in the same area. Although they are not included in these measurements, ellipsoidal mafic schlieren and primary foliations in U1f parallel the sheets as well.

In some parts of the HZ regular, well-oriented structures are absent - notably near the margins of the pluton where xenolithic blocks are abundant and near the HG-HZ contact, and locally in other areas as well.

Unit 2

Abundant, discrete mafic pillow trains (U2m) are hosted by granitic rocks (U2f) in sheets that cross-cut U1 rocks (Fig. 8). Some of these zones have sharp boundaries and are clearly composite dikes. In others the sheet geometry is defined entirely by the pillow swarm and the host granite does not have sharp, parallel, well-defined walls. The shapes of the ellipsoidal pillows define a fabric that is parallel to the sheet. U2 sheets are uniformly steeply dipping and north-striking and are typically ~10-15 m across. Individual pillows are about 0.5-2 m in length and typically have crenulate contacts with U2f. Some of the U2m pillows are partially bounded by fine-grained intermediate rock that in turn is in sharp contact with U2f rock.

U2m rocks are finer grained and generally darker in color than most U1m rocks. They range in composition from diabase to quartz monzodiorite; plagioclase and hornblende are the most abundant minerals, biotite and K-feldspar are present in lesser abundance, and most contain minor quartz. Clinopyroxene is present as inclusions in plagioclase or relict cores in hornblende in most samples, and pseudomporphs after clinopyroxene and probably olivine are common. Apatite is acicular.

The granites of U2f are medium-grained, with prominent subhedral feldspars and interstitial quartz. Plagioclase-mantled K-feldspar is common (rapakivi texture). Some U2f rocks, especially those with prominent rapakivi texture, have a porphyritic texture with abundant feldspar phenocrysts and a fine-grained phaneritic groundmass. Small miarolitic cavities are common, especially in more felsic rocks. U2f has a wide range of mafic mineral content (biotite, hornblende, minor clinopyroxene totalling 5-30%). Where mafic minerals are more abundant, they commonly cluster and display reaction textures, suggesting that they represent incorporation of mafic material from U2m. Euhedral sphene is abundant, and apatite is strongly acicular. Zircon and allanite are also readily apparent. The less mafic U2f rocks are almost identical to granites of the HG unit, except that HG lacks rapakivi texture. U2f differs from U1f in that it is characterized by more felsic compositions, higher percentages of k-feldspar, and rapakivi textures.

Heterogeneous Zone-Homogeneous Granite Contact

The HZ-HG contact has a roughly E-W average trend, but it is highly irregular (Fig. 1). Long (~1-2 km) NNE/SSW sections of the contact east of Tule Wash are normal faults, with HG dropped down to the west and juxtaposed against HZ. Somewhat smaller irregularities (up to ~1 km) are primary magmatic features. In one area east of Tule Wash fault, a chain of 1 - 100 m discrete masses of HZ mafic rock extends 1 km into HG. Elsewhere, the contact is very well defined. Typically, in unfaulted areas, abundant large pillows of the HZ decrease gradually in abundance northward within a 50-100 m wide HZ-HG transition zone. The pillows here are hosted by U2f granite with very well developed rapakivi texture and relatively fine grained groundmass. The porphyritic rapakivi texture continues beyond the last exposure of mafic rock and gradually fades as the granite eventually assumes typical HG texture. Where it is well exposed in three dimensions, the contact dips gently northward (Fig. 9).

Late Dikes

Abundant, north-south trending, sub-vertical dikes crop out throughout the HZ and extend across the contact into the HG. They are most abundant west of Tule Wash fault and in the north-central part of the HZ east of the fault. Some or all of these dikes belong to the Ireteba dike swarm, which extends southward for several km into country rock [8]. The 2 - 8 m thick dikes include aphanitic mafic porphyries, granites and granite porphyries (commonly with rapakivi texture), and some fine-grained to aphanitic intermediate rocks. Large pillows are absent, but many of the felsic dikes have small mafic enclaves and/or mafic margins.

The margins of the dikes are typically finer grained than the interiors. Some dikes have undulating, locally gradational contacts with mafic and felsic rocks from Units 1 and 2.

INTERPRETATION OF MAGMATIC STRUCTURES

The distinctive structures preserved within the HZ provide clear evidence for repeated interaction between contrasting magmas [cf. 15]. Most obviously, ubiquitous mafic pillows with delicate quenched, crenulate margins against more felsic rock indicate chilling of mafic magma against felsic magma. Widespread disequilibrium and quenching textures, such as mantling of K-feldspar by plagioclase in U2f granite, mafic clots with altered pyroxene cores in felsic rocks, abundant acicular apatite, and abrupt strong compositional discontinuities in plagioclase also attest to open system processes [8]). We discuss below our further interpretation of the features that are present within the HZ.

Structures in Unit 1
Sheets and Pipes: Mafic Ponding, Felsic Entrapment and Ascent, and Implications for Original Orientation

The most prominent structures in U1m strongly resemble those described by Wiebe [27,28] from complexes on the coast of Maine, and by Snyder and Tait [25] for experimental interaction of fluids of contrasting density. Specifically, the distinctive mafic sheets and intervening relatively mafic granitic rocks and the sheet- and pipe-like felsic protrusions are all virtually identical to features described by Wiebe. For the most part, we follow his interpretation of these features in the discussion below.

The mafic sheets represent mafic magma that intruded into an active felsic magma chamber. At the base of the chamber, the mafic magma may have diked upward through relatively dense, rigid, crystal-rich material, but ultimately when it intruded into less dense felsic magma it began to flow laterally toward the magma chamber walls, ponding in horizontal sheets. Felsic rock between the sheets was trapped beneath the overlying sheet, perhaps as "accidental" pods of lower density magma trapped between lobes, or possibly as crystal-rich cumulate near the bottom of the active portion of the felsic chamber over which the mafic sheet spread. Coarse feldspars within mafic sheets may be derived from the felsic sheets, having been incorporated by the hardening, cracking edges of the mafic magma as it flowed over the felsic mush.

The trapped sheets of felsic material would have been inherently unstable gravitationally. If they represent trapped felsic melt zones, they would initially have been less dense than the overlying mafic material. If they represent initially crystal-rich material over which mafic melt flowed, they may at first have had comparable density to the overlying sheet. However, as the felsic material warmed and became more molten and the mafic sheet cooled and solidified, a density inversion would have resulted. Like Wiebe [27,28], we interpret the pipe-like and sheet-like protrusions to represent upward injections of less-dense felsic magma into the overlying mafic sheets. The very felsic granite in these structures was probably crystal-poor, whereas the "trapped granite" between mafic sheets was cumulate material. The relatively mafic, quartz-poor, high Sr, plagioclase-rich composition and the orientation of plagioclase laths parallel to the mafic sheets are consistent with cumulate origin of the intervening felsic sheets. Crystals may have accumulated from overlying granitic magma, prior to emplacement of the mafic sheet, or they may be residues after extraction of melt that was squeezed off to form protruding sheets and pipes and perhaps larger masses at higher levels.

Because they are gravitationally controlled, the attitudes of these magmatic structures can be used to reconstruct the original orientation of the pluton [e.g. 25,27,28]. Mafic sheets would have ponded horizontally, and magma that subsequently ascended from trapped felsic layers should have ascended vertically to form the pipes and protruding sheets. We therefore interpret the orientations in Fig. 1c to indicate the magnitude and direction of tilting in different parts of the pluton - generally about 20° NE east of the Tule Wash fault, and 20° WNW west of the fault. This would indicate that the top of the magma chamber is represented by the northern, HG section of the pluton, and that the HZ represents the base of the magma chamber. This is to be expected because dense mafic magma would pond at the base of a felsic magma chamber.

Local areas with chaotic orientations of pillows and enclaves may represent zones of continued activity that disrupted the initial structure. This could have been a result of entrapment of thick layers of granite beneath mafic sheets (hence massive reintrusion), or of disruption of layers by continued mafic injections.

Coarse Mafic Rocks: Cumulates

We interpret the coarse-grained, highly mafic rocks to be cumulates, as suggested by both their textures and compositions [cf. 8]. Commonly, they are rhythmically layered. Their layering is subparallel to nearby mafic sheets, suggesting that they were deposited horizontally (Fig. 1c). Cumulates probably formed from very thick mafic sheets which cooled slowly, permitting growth and deposition of coarse crystals. Collapse of thick, high-density cumulate-rich sheets may have resulted in downward dispersal of angular blocks that are now found hosted by felsic rock and associated with more typical fine-grained pillows (see Fig. 4b).

Xenolith Zones: Disaggregation of Wall Rock, Interruption of Sheet Emplacement

We suggest that country rock now represented by the abundant xenoliths near the margins of the HZ was intensely injected by early dikes to the point of disaggregation. As blocks of country rock became bounded by dikes, the walls lost coherence and collapsed into the pluton (note abundant early dikes cutting blocks - Fig. 7). Within this zone, the absence of oriented mafic sheets may reflect either interference with spread of the sheets by the abundant blocks, or collapse of the margin of the chamber relatively late in its history, after intrusion of much of U1.

Structures in Unit 2

Unit 2 represents a fundamental change in the style of intrusion, from dominantly lateral flow to vertical flow. The diffuse contacts between many U2f rocks and surrounding U1 suggest that the HZ was at least partly molten when this change occurred. The uniform N strike of the U2 sheets may indicate that the pluton was sufficiently rigid to transmit the regional ~E-W extensional stress [cf. 8,11,13].

The abundant pillows that define the character and orientation of U2 sheets suggest a fresh input of mafic magma into the chamber. It is highly unlikely that already solidified pillows were entrained by ascending granitic magma. Large mafic pillows with delicate crenulate margins typically comprise ~50% or more of Unit 2 sheets; these pillows could not all be accidental solid inclusions incorporated from the walls of a granite dike. Rather, we envision U2 as a very coarse slurry of mafic and felsic magma. Influx of mafic magma may have been accompanied by fresh influx of felsic magma; it may have remelted and mobilized largely solid felsic material near the base of the magma chamber; or it may have facilitated ascent of granitic magma that had been trapped beneath U1 at the base of the chamber. We favor the latter interpretation.

Emplacement of U2 may have entirely postdated U1 and represent a completely separate injection episode. Alternatively, U2 sheets may mark vertical conduits that fed overlying, horizontal U1m sheets.

The Heterogeneous Zone-Homogeneous Granite Contact

The HZ-HG contact was broadly subhorizontal prior to tilting, with denser material below and less dense above, and it is gradational on a scale of 10's of m. These characteristics demonstrate that it does not represent intrusion of molten magma into already solid rock. However, the facts that the contact has hundreds of meters of steep relief and that an initially vertical train of mafic masses is preserved within the HG indicate that the HG granite had significant strength when the uppermost part of the HZ was emplaced. Otherwise, this gravitationally unstable configuration would not have been preserved. The upper part of the HZ appears to be entirely made up of U2 rocks that have an atypically chaotic structure (not in discrete sheets; no preferred orientation of pillows). We suggest that HG was semisolid at the time of final intrusion of the composite U2 magma and had sufficient strength to permit diking by U2 and to prevent complete collapse of U2 protrusions. Nontheless, the relatively low density of HG presented a barrier to continued rise of U2 magma, and so for the most part the U2 slurries spread laterally at the HZ-HG contact. The fact that U2 magma was composite prevented development of well-defined sheets. Ascent and partial collapse of columns of ascending U2 may have added to the chaotic orientations near the HG-HZ contact.

Late Dikes

The late dikes postdate all of the other units and structures of the Aztec Wash pluton. We concur with Falkner et al. [8] that many and perhaps all of these dikes were emplaced very late in the active history of the magma chamber, when it was entirely rigid and affected by the regional E-W extension, but at least locally contained some melt. The mafic dikes, and quite likely the felsic dikes as well, represent fresh influxes of magma into the pluton.

RECONSTRUCTING THE PLUTON

Map patterns and magmatic structures permit reconstruction of the initial geometry of the pluton. The cross section is especially important, because the thickness of the pluton as a whole and of individual units strongly influences thermal structure and related characteristics such as crystallization rates and magma viscosity, which in turn control convection and potential longevity [e.g. 23, 26].

Structures that indicate horizontal orientation at the time of intrusion (ponded mafic sheets, felsic pipes and sheet-like protrusions, possibly cumulate layering) permit restoration of pre-tilt geometry, as discussed above. We can estimate thickness of the pluton by using the average dips map patterns, and elevations of the contacts. East of the Tule Wash fault, dips average $22°$ to the NNE, the NNE extent of the pluton is ~6.5 km, and the difference in elevation between the NE and SW contacts of the pluton is negligible. These data yield an estimate of ~2.4 km for the vertical dimension of the pluton. The HZ averages about 1.8-2.0 km of this thickness, and the HG about 0.4-0.6 km. (For a more thorough treatment of estimation of the initial thickness, see [20).

As a test of our reconstruction of the thickness of the pluton, we selected seven samples from the southern and middle portion of the HZ for hornblende barometry [20], complementing analyses from four locations at the northern margin of the HZ by Falkner [8]. Appropriate assemblages are absent from the HG. We rejected two of our samples for inclusion in interpretation because one had too mafic a composition for application of the barometer (no quartz in contact with hornblende), and the amphibole in the other was highly actinolitic (altered?). We calculated temperatures for samples (including Falkner's) using the plagioclase-hornblende thermometer of Blundy and Holland [4]. Average temperature estimates for our five samples and Falkner's four samples, based on plagioclase and hornblende rim compositions, ranged from 650 to 685°C, presumably reflecting equilibration at or very near the vapor-saturated solidus. For pressure estimates, we used the method of Anderson and Smith [2], which is based on experimental results of Johnson and Rutherford [16] and Schmidt [24] and incorporates the effect of temperature.

Pressure estimates for our two samples collected near the southern margin (bottom of pluton) average 2.1 kb; the three from the middle of the HZ average 1.7 kb; and Falkner's four from the northern HZ average 1.5 kb. The difference between bottom and top of HZ of about 0.6 kb is equivalent to about 2 km difference in depth, which is consistent with our estimate of the thickness of the HZ.

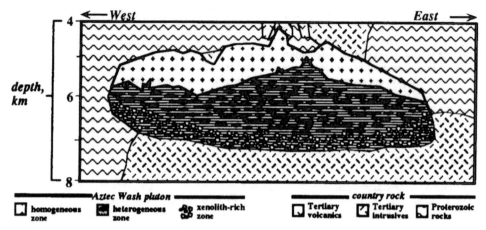

Figure 10. Schematic cross-section of pre-tilting, pre-faulting Aztec Wash pluton, based upon map relations and tilt estimates (see text).

Figure 10 illustrates our reconstruction of an east-west cross section of the pluton, restored to pre-tilting and pre-fault displacement configuration. Overall, the pluton is lensoidal with an aspect ratio of about 4.5. The thick HZ dominates the intrusion and underlies the thinner HG, which in its homogeneity preserves no obvious record of the pluton's complex history.

DISCUSSION

History of the Aztec Wash Magma Chamber

The following history can be established for Aztec Wash pluton, based primarily on the structures within the HZ:

(1) A thin granitic magma chamber (HG), on the order of 1 km thick, formed within the upper crust about 5 km below the surface. It began to crystallize and may have accumulated a relatively thick crystal-rich mush at its base

(2) Repeated pulses of mafic magma (U1m) injected into the magma chamber and ponded, trapping granite (U1f; perhaps felsic crystal mush) beneath them and cooling as they spread subhorizontally toward the walls. Thicker mafic sheets cooled slowly enough to accumulate coarse crystals. The gravitational instability of the trapped felsic sheets resulted in remobilization, with extracted felsic melt intruding overlying mafic sheets to form pipes and sheet protrusions. Melt-depleted cumulate remained within the trapped sheets. Some hybridization between felsic and mafic magmas, via both mechanical and diffusive mixing, occurred during and after emplacement of the mafic sheets.

(3) During emplacement of U1, country rocks were heavily injected and failed, forming xenolith-rich marginal zones.

(4) Fresh pulses of ascending mafic magma (U2m) injected into the magma chamber, activating felsic magma (U2f) that was trapped beneath or within U2. Coarse slurries of the mafic and felsic magmas diked through the overlying U1. U2 dikes may represent a discrete intrusive episode, or they may have been feeders for high-level U1 sheets. In either case, fracturing of the still partly molten U1 to permit ascent of U2 may have been facilitated by regional stresses related to the initiation of extension. By the time of final injection of U2, the granite of HG was also sufficiently crystalline to have signicant strength, as indicated by the irregular upper surface of the HZ and by U2 injections preserved within HG.

(5) Late mafic and felsic dikes intruded after the entire pluton was very nearly, but apparently not quite, solid. These dikes, like U2, appear to have filled fractures whose formation was facilitated by regional extension.

(1-5?) Throughout its history, the Aztec Wash magma chamber may have fed eruptions [8]. Voluminous volcanic strata of appropriate age are abundant in the area. Persuasive evidence that eruptions tapped this chamber is lacking, however.

Implications for Longevity of the Magma Chamber

A magma chamber of this size, at this shallow depth, would not remain above its solidus long if it were formed in a single episode by injection of a single magma. However, recharging of an initially felsic magma chamber by much hotter mafic magma could could provide an enormous heat input that would at least in part offset conductive heat loss to the country rock and thus maintain at least some of the felsic magma above its solidus for a protracted period. The effect would be much greater than for recharging a mafic magma chamber with mafic magma, or a felsic magma chamber with felsic magma. Mafic magma would probably be at least 400° above the solidus of granite, in contrast to 100° to, at most, 200° that magmas are above their own solidi when they are intruded. Frequent injections of mafic magma into a felsic chamber could thus greatly prolong the lifetime of granite as a crystal mush compared to interval that it would survive without the injections.

It is not clear whether, during its complex history, the Aztec Wash magma chamber was ever entirely solid prior to final solidification. It is certain, however, that material in the pluton that hosted injections was partially molten (some melt present, but in most cases significant strength) during most or all of the intrusive events we have identified. This could indicate that (a) host material had solidified and was partially remelted by intrusions; (b) intrusions were closely spaced in time so that they all occurred during the crystallization interval of the host; or, an intriguing variant of (b): (c) repeated injections balanced heat loss to country rock, so that the crystallization interval was greatly prolonged.

We do not yet have any quantitative estimate of the duration of the events that are recorded in Aztec Wash pluton. We intend to pursue this question through a combination of thermal modelling, detailed geochronology, and studies of minerals as recorders of the nature and duration of magma chamber events.

Utility of Magmatic Structures in Unravelling Geometry of Tectonic Structures

Because they provide an indication of paleohorizontal and therefore of post-pluton tilting, the structures that we describe here are potentially very useful in tectonic investigations [cf. 28]. The structural evolution of the central Eldorado Mountains provides a case in point. Oppositely dipping volcanic and sedimentary strata to the north (E-dipping) and south (W-dipping) of Aztec Wash pluton indicate the presence of an antiformal accommodation structure in the central part of the range [e.g. 9-11]. However, stratified rocks are absent here, so documenting the precise geometry of the accommodation zone is difficult. The structures in Aztec Wash provide data illuminate this problem: the opposing tilts on opposite sides of the Tule Wash fault (NNE dipping on the NE, WNW dipping on the SW) indicate that hinge of a NM-plunging antiform predicted by Faulds must run near the fault [10,22]. Similar applications can be made elsewhere where such magmatic structures are present and bedded strata of appropriate age are not.

ACKNOWLEDGEMENTS

Claudia Seifert-Falkner and Robert Wiebe were instrumental in making this study possible. Seifert-Falkner's excellent investigation of the Aztec Wash pluton formed the basis from which this study grew. Wiebe's work on magmatic structures provided a model for our project, both for specific interpretations and for ways of pondering and investigating them. His field trip to the coast of Maine in 1995, in connection with the Hutton Symposium, was inspirational. In addition, Jim Faulds, Jonathan Miller, Rod Metcalf, John Ayers, Carolyn Bachl, Delores

Robinson, Alan Wiseman, Leonard Alberstadt, Allen Glazner, Art Sylvester, John Bartley, and Barbara John all contributed in various ways to this project. Support was provided by NSF grants EAR 92-19860 and 96-28380 and grants from Sigma Xi, the Massachusetts Teachers' Association, and the Vanderbilt Department of Geology.

REFERENCES

1. J.M. Amato, J.E. Wright, P.B. Gans, and E. Miller, Magmatically induced metamorphism and deformation in the Kigluaik gneiss dome, Seward Peninsula, Alaska, *Tectonics*, 13, 515-527 (1994).
2. J.L. Anderson and Smith D.R., The effects of temperature and f_{O2} on the Al-in-hornblende barometer, *Amer. Min.*, 80, 549-559 (1995).
3. R. Bateman, The interplay between crystallization, replenishment and hybridization in large felsic magma chambers, *Earth Science Reviews*, 39, 91-106 (1995). .
4. Blundy, J.D., and T.J.B Holland, Calcic-amphibole equilibria and a new amphibole - plagioclase geothermometer, *Contrib. Minel . Petrol.*, 104, p.208-224 (1990).
5. D.S. Coleman, A.F. Glazner, J.S. Miller, K.J. Bradford, T.P. Frost, J.L. Joye, and C.A. Bachl, Exposure of a Late Cretaceous layered mafic-felsic magma system in the central Sierra Nevada batholith, California, *Contrib. Mineral. Petrol.*, 120, 129-136 (1995).
6. J. Didier and B. Barbarin (eds.), *Enclaves and granite petrology*, Elsevier, Amsterdam (1991).
7. Falkner, C.M., *History and petrogenesis of Aztec Wash pluton, Eldorado Mountains, Nevada* [unpublished M.A. thesis], Vanderbilt Univ., Nashville, Tennessee (1993).
8. C.M. Falkner, C.F. Miller, J.L.Wooden, and M.T. Heizler, Petrogenesis and tectonic significance of the calc-alkaline, bimodal Aztec Wash pluton, Eldorado Mountains, Colorado River Extensional Corridor, *Jour. Geophys. Res.*, 100, 10,453-10,476 (1995).
9. J.E. Faulds, J.W. Geissman, and C.K Mawer, Structural development of a major extensional accommodation zone in the Basin and Range Province, northwestern Arizona and southern Nevada: Implications for kinematic models of continental extension, *Geol. Soc. Am. Memoir* 176, 37-76 (1990). .
10. J.E. Faulds, J.W. Geissman. and M. Shafiqullah, implications of paleomagnetic data on Miocene extension near a major accommodation zone in the Basin and Range Province, northwestern Arizona and southern Nevada, *Tectonics*, 11, 204-227 (1992).
11. J.E. Faulds, New insights on the geometry and kinematics of the Black Mountains-Highland Spring Range accommodation zone (BHZ), Arizona and Nevada, *Geol. Soc. Am. Abstr. Programs*, 26, 51 (1994).
12. T.P. Frost and G.A. Mahood, Field, chemical, and physical constraints on mafic-felsic magma interaction in the Lamark Granodiorite, Sierra Nevada, California, *Geol. Soc. Am. Bull.*, 99, 272-291 (1987).
13. P.B. Gans, G.A. Mahood, and E. Schermer, Synextensional magmatism in the Basin and Range Province; A case study from the eastern Great Basin, *Geol. Soc. Am. Special Paper* 233, 53 p. (1989).
14. P.B. Gans, Extensional strain rates in the Basin and Range Province, *Geol. Soc. Am. Abstr. Programs*, 23, 44 (1991).
15. K.A. Howard and B.E. John, 1987, Crustal extension along a rooted system of continental extensional tectonics, *Geol Soc. London Spec. Pub.*, 28, 128-176
16. M.C. Johnson and M.J. Rutherford, Experimental calibration of the aluminum-in-hornblende geobarometer with application to Long Valley caldera (California) volcanic rocks, *Geology*, 17, 837-841 (1989).
17. R.V. Metcalf, E.I. Smith, J.D. Walker, R.C. Reed, and D.A. Gonzalez, Isotopic disequilibrium among commingled magmas: evidence for a two-stage magma mixing-commingling process in the Mt. Perkins pluton, Arizona, *J. Geol.*, 103, 509-527 (1995).
18. P.J. Michael, Intrusion of basaltic magma into a crystallizing granitic magma chamber: The Cordillera del Paine pluton in southern Chile. *Contrib. Mineral Petrol.*, 108, 396-418 (1991).
19. C.F. Miller, C.A. Bachl, J.S. Miller, .JL. Wooden,, J.E.Faulds, and M.L. Shaw, Mid-crustal plutons of the Eldorado Mountains: evidence for large-scale magmatic modification and reorganization of the crust in the Colorado River extensional corridor, *Geol. Soc. Am. Abstr. Programs*, 27, A435 (1995)
20. D.W. Patrick, 1996, *Implications of the style of intrusion into a composite, recharging magma chamber: The Aztec Wash pluton* [unpublished M.A. thesis], Vanderbilt Univ., Nashville, Tennessee (1996).

21. Patrick, D.W., Miller, C.F., and C.F. Falkner, The Aztec Wash pluton: Implications for the style of intrusions into a composite, recharging magma chamber, *30th International Geological Congress Abst. with Prog.*, 2, 411 (1996).

22. D.W. Patrick, C.F. Miller, and J.E. Faulds, Magmatic structures in Aztec Wash pluton: implications for extension in the Eldorado Mountains, Nevada, *Geol. Soc. Am. Abstr. Programs*, 28, A512 (1996).

23. Pitcher, W.S., *The nature and origin of granite*, Blackie Academic and Professional, London (1993.)

24. M.W. Schmidt, Amphibole composition in tonalite as a function of pressure: an experimental calibration of the Al-in-hornblende barometer, *Contrib. Mineral Petrol.*, 110, 304 - 310 (1992).

25. D. Snyder and S. Tait, Replenishment of magma chambers: comparison of fluid-mechanic experiments with field relations, *Contrib. Mineral. Petrol.*, 122, 230-240 (1995.)

26. R.S.J. Sparks and L.A. Marshall, Thermal and mechanical constraints on mixing between mafic and silicic magmas, *J. Volc. Geoth. Res.*, 29, 99-124 (1986).

27. R.A. Wiebe. The Pleasant Bay layered gabbro-diorite, Coastal Maine: Ponding and crystallization of basaltic injections into a silicic magma chamber, *Jour. Petrology*, 34, 461-489 (1993).

28. R.A. Wiebe, Silicic magma chambers as traps for basaltic magmas: The Cadillac Mountain intrusive complex, Mount Desert Island, Maine, *Jour. Geology*, 102, 423-437 (1994).

Proc. 30th Int'l. Geol. Part 15 pp. 137–152
Li et al. (Eds)
© VPS 1997

Fractional Crystallization, Magma Mixing and Crustal Assimilation in the Evolution of Plateau and Rift Magmatism in Ethiopia

ANGELO PECCERILLO
Dipartimento di Scienze della Terra, 87036 Arcavacata di Rende (Cs), Italy

GEZAHEGN YIRGU, MEGERSSA BEKELE
Department of Geology and Geophysics, Addis Ababa University, PO Box 1176, Addis Ababa, Ethiopia

TSAI WAN WU
Department of Geology, University of Western Ontario, London, Ontario, Canada

Abstract

Magmatic activity is widespread in Ethiopia. Oligocene to Pliocene flood basalts, ignimbrites and shield volcanoes cover an area of several hundred thousand square km. Recent to active basaltic to rhyolitic volcanism occurs within the Ethiopian Rift Valley and the Afar depression. Major and trace element studies, together with preliminary Sr isotopic investigations indicate that rhyolitic magmas from both the plateau and rift sequences have been formed by Fractional Crystallization (FC) and Assimilation plus Fractional Crystallization (AFC) processes, from basaltic parental magmas. Both plateau and rift basalts have undergone complex evolutionary processes, which have substantially modified their trace element composition. Mixing between basaltic magmas and associated acid liquids is responsible for the geochemical modifications of plateau and rift basalts. Hybrid mafic magmas display variable ratios of elements such as Rb/Ba, Rb/Zr, etc., which have an incompatible behaviour in mafic systems. Interaction between acid and mafic melts is also indicated by the widespread occurrence of xenocrystal phases (e.g. sanidine, quartz, etc.) in several basaltic rocks. Bulk crust assimilation processes have affected rift basalts.

The present data clearly indicate that many mafic magmas in the Ethiopian plateau and rift valley do not represent primary melts and, in most cases, their composition has been significantly modified by evolutionary processes. This places serious problems to the identification of mantle reservoirs and stresses the need for detailed investigations on single suites in order to discriminate between source and shallow level geochemical signatures.

Keywords: magmatism, flood basalts, rift volcanism, petrology, geochemistry, Ethiopia, Cenozoic, Quaternary

INTRODUCTION

Continental breakup is often associated with widespread magmatic activity [e.g., 25] The magmas generally display important temporal compositional variations, which may be related either to changes in the style of magma evolution and/or to tapping of compositionally different mantle reservoirs. This case is met in Ethiopia, where magmatic activity of variable composition has been going on from the Oligocene to the Present time. [17, 21, 3, 18]. The volcanism is associated with continental breakup and with the formation of the East Africa Rift Valley. The mafic magmas display variable composition, from tholeiitic to transitional and alkaline [21].

According to some authors the petrologic and geochemical variations of basalts in Ethiopia reflect variable mantle sources beneath East Africa rift zone [e.g. 10]. Other authors suggest that different degrees of partial melting of a single mantle reservoir may be responsible, at least partially, for the observed geochemical differences in basaltic magmas [e.g. 21]. Such a problem is not restricted to East Africa, but has been debated in most if not all volcanic provinces occurring on continental crust [e.g. 12, 5, 16].

The present paper is aimed at investigating this topic by discussing geochemical data in three suites of Ethiopian basaltic lavas and associated rhyolites from both plateau and rift environments. It will be shown that fractional crystallization, mixing and assimilation processes have important effects in generating the compositional spectrum of the studied series. These processes had an important role in modifying the geochemical characteristics of both rift and plateau basalts.

GEOLOGICAL SETTING

The Ethiopian volcanic province covers an area greater than 600.000 km^2. It is dominated by fissure-fed basaltic lavas forming the volcanic plateaus that bound the Afar and Ethiopian Rift [18]. A number of studies [e.g. 26, 17, 10, 3, 14] allow distinction of two main magmatic stages. During the first stage, Oligocene to Pliocene large fissure eruptions of basalts built up thick flood lava sequences (known as Ashange and Aiba Basaltic Formations), associated with late ignimbritic sheets (Alaji Rhyolitic Formation). This magmatic stage was closed by the formation of huge basaltic shield volcanoes (Termaber Basalt Formation). K/Ar age determinations reported by several authors [see 17, 18 for review] show a maximum emplacement age of about 55 Ma for the Ashange basalts, an age range of 34 to 13 Ma for the Alaji Formation and an interval of 28 to 5 Ma for the Termaber Formation. New $^{40}Ar/^{39}Ar$ age determinations are changing the earlier picture and indicate that the flood basalt volcanism in the Central Ethiopian Plateau region came to an end at about 30 Ma and was probably emplaced in a shorter time interval than previously envisaged [11].

The second stage of volcanic activity is Pliocene to Recent in age and is more closely related to the formation of the Rift Valley and Afar [6, 9, 2]. In the Ethiopian Rift Valley, volcanic rocks consist essentially of rhyolitic and trachytic plinian pumice and ignimbrites with minor lava flows. These are associated with volumetrically subordinate basaltic cinder cones and lava flows that are aligned along extension faults parallel to the rift. Intermediate rocks are very scarce. Acidic products are often associated with large central composite volcanoes with summit calderas, such as at Gedemsa, Gariboldi, Fantale [9, 19].

All the volcanic products rest either directly over the metamorphic rocks of the Precambrian basement, or over the Mesozoic sedimentary sequence.

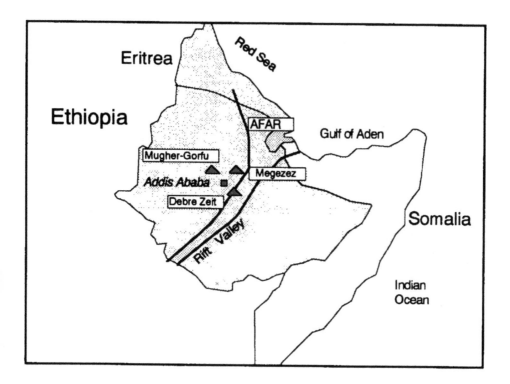

Figure 1. Location map of the studied volcanic suites

SAMPLING AND PETROGRAPHY

The samples studied in this work include two rock series from the first stage activity, and one suite from the border of the rift valley. Samples come from the Mugher river gorge and Gorfu hills, from the Megezez volcano near the town of Debre Birhan, and from Debre Zeit. The sampling localities are shown in Fig. 1.

The Gorfu-Mugher sequence [20] consists largely of basalts and one ignimbrite sheet. The ignimbrite is interbedded between the Mugher (plateau basalts) and Gorfu (central volcano) products. The K/Ar age of this sequence ranges from 28 to 5 Ma [17]. The Megezez volcanics include mafic to acid rocks and belong to the Alaji and Termaber formations. The age is apparently younger than about 5 Ma [17, and references therein]. The Debre Zeit basalts are Quaternary, with some eruptions occurred in historic times [8].

The basaltic rocks have a texture ranging from aphyric to strongly porphyritic. Several rocks, especially those from the first stage activity, show phenocrysts and megacrysts of

Table 1. Composition of representative volcanics from Mugher-Gorfu, Megezez and Debre Zeit. Asterisks indicate data from Gasparon et al. [8].

Sample	Mugher-Gorfu				Megezez					Debre Zeit				
	ETP-1	ETP-2	ETP-3	ETP-4	MEZ-1	MEZ-2	MEZ-3	MEZ-4	MEZ-5	AA34*	AA56	Y5*	AA55*	AA6*
SiO_2	47.20	48.19	50.13	67.18	46.76	49.55	60.59	67.81	68.53	48.37	50.38	57.46	72.78	74.92
TiO_2	1.13	2.03	2.57	.47	2.39	3.23	1.36	.63	.68	1.92	1.71	1.96	.25	.06
Al_2O_3	16.99	15.25	16.39	10.01	18	14.99	15.22	12.89	15.27	16.73	17.20	15.57	9.36	12.33
$FeOt$	10.76	10.85	10.91	6.59	11.43	12.35	7.10	4.23	5.90	10.25	8.40	9.39	5.55	0.53
MnO	.18	.15	.13	.23	.16	.18	.17	.17	.22	.16	.16	.22	.19	.06
MgO	9.63	6.73	4.50	.26	3.91	3.58	1.19	.44	.32	7.99	6.99	1.32	.03	.10
CaO	8.88	9.38	8.56	.54	8.62	6.84	3.98	.33	.32	9.68	9.00	4.37	.25	.48
Na_2O	2.73	2.74	3.29	4.16	3.08	3.46	4.64	3.62	4.67	2.09	3.16	4.54	6.02	4.21
K_2O	.50	1.14	1.44	3.84	.95	1.58	2.93	4.41	4.26	1.17	1.86	2.79	4.04	4.45
P_2O_5	0.31	.57	.54	.04	.45	.65	.58	.6	.8	.46	.44	.9	.01	.01
V	200	259	263	11	304	268	22	8	8	272	219	87	3	4
Cr	370	155	30	1.01	34	-	-	-	-	408	181	2	1	1
Sc	29	27	24	2.05	19	22	8	6	6.07	31	26	19	2	7
Co	43	43	33	1	42	33	4.7	1.4	2.7	60	38	14	2	1
Ni	152	66	13	2	37	8	7	10	9	422	84	5	2	1
Rb	6	21	33	141	16	32	80	113	123	21	49	62	216	236
Sr	319	533	626	15	647	497	600	14	32	558	594	549	1	10
Ba	206	516	466	139	298	560	756	290	152	445	604	1270	24	18
Y	26	32	37	101	33	70	63	222	143	26	30	74	163	49
Zr	85	199	227	1206	165	305	379	910	1275	166	192	398	1339	129
Nb	9	33	33	161	18	37	43	123	147	37	51	70	144	93
Ta	.6	1.07	1.08	7.04	1.0	2.0	2.0	5.1	8.1	2.03	3.05	4.09	15	7.04
Hf	2.1	4.0	6.0	24	3.1	7.1	8.0	19	28	4.04	5.03	9.01	29	6.01
La	9.1	31	34	167	18	46	53	223	132	31	41	86	185	14
Ce	23	63	69	293	38	90	105	297	274	55	71	142	349	35
Nd	11	30	34	162	20	54	53	188	109	22	29	75	132	21
Sm	2.1	6.1	6.1	28	4.1	12	11	32	20	6.3	7.2	17.5	28	5.2
Eu	1.0	2.0	2.0	2.1	1.1	4	3.0	5.1	4	1.97	2.2	4.5	1.5	.17
Tb	.5	.8	.9	3.1	.7	1.1	1.1	5.0	3.0	1.2	1.3	2.1	3.7	5.6
Yb	1.1	1.1	2.0	11.1	1.1	4.0	3.1	15	11	2.4	2.4	5.6	15	5.6
Th	.9	3.0	3.0	18	1.1	3.0	8.0	13	21	2.8	6.7	8.2	29.2	23.5
$^{87}Sr/^{86}Sr$										.70388	.70372	.70452	.70762	.70569

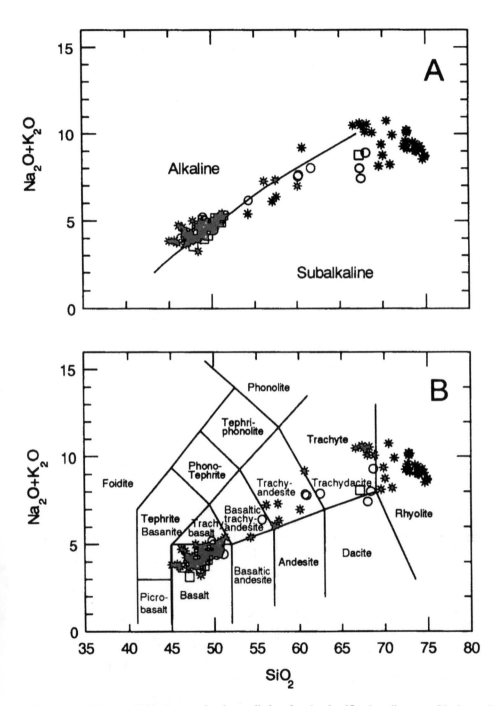

Figure 2. Alkalies vs. SiO$_2$ diagram for the studied rocks. A: classification diagram of Irvine and Baragar [13]. B: classification grid of Le Bas et al. [15]. Symbols: asterisks: Debre Zeit; open squares: Mugher-Gorfu; open circles: Megezez.

plagioclase and clinopyroxene and phenocrysts of olivine set in a holocrystalline to hypocrystalline groundmass. Plagioclases are strongly zoned and sometimes corroded. Olivine occurs as euhedral to subhedral crystals sometimes transformed to iddingsite and to other secondary products. Olivine is the dominant phenocrystal phase in the Debre Zeit basalts. Pyroxene generally occurs as euhedral crystals. Magnetite is an ubiquitous phase in the groundmass and, sometimes, as a microphenocryst. Groundmasses consist of the same phases as the phenocrysts and of opaque minerals; textures range from intersertal to microphytic. Altered glass is observed in some samples. Xenocrysts of corroded quartz and sanidine are often observed.

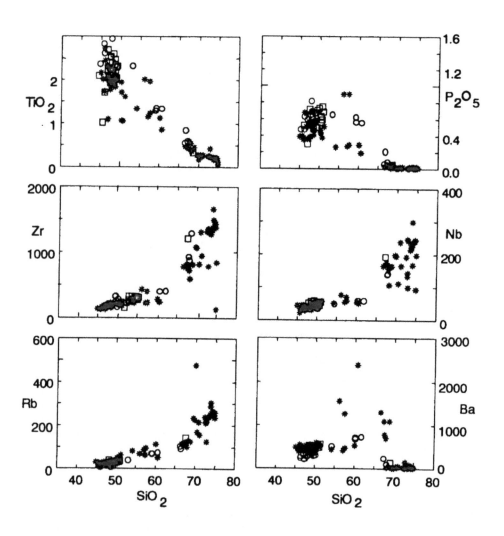

Figure 3. Variation diagrams of some major and trace elements. Symbols as in Fig. 2.

Acid rocks are poorly porphyritic with a few phenocrysts or fragments of phenocrysts of alkali feldspars and corroded quartz set in a glassy or cryptocrystalline groundmass. Green pleochroic clinopyroxene and fayalite are often present in small amounts.

GEOCHEMISTRY

Analytical methods

The present paper discusses both literature data and newly determined major and trace element contents. Composition of representative samples are reported in Table 1. Major elements have been analysed by combined wet chemical methods and X-ray fluorescence procedures [7]. The trace elements V, Ni, Rb, Sr, Y, Nb, Ba and Zr have been analyzed by XRF. Several international rock sta*ndards have been used for curve calibration. Precision is better than 10% for all the trace elements, except Rb and Sr, which have precisions better than 5%. REE, Sc, Cr, Co, Th, U, Cs, Ta and Hf have been determined by INAA at the Department of Geology, University of Western Ontario. Precision is better than 10% for most of these elements at the concentration level of the studied samples.

Major and trace elements

Variation diagrams for major and some trace elements are reported in Fig. 2, 3. TAS classification diagram [15] shows that the studied suites of rocks vary from basalts to rhyolites. The basalts plot along the boundary separating the alkaline and the subalkaline series according to Irvine and Baragar [13] (Fig. 2A) and are best characterized as transitional basalts, following the nomenclature of Piccirillo et al. [21] for Ethiopian volcanics. Basalts do not show important time-related geochemical variations for many of the considered geochemical parameters, except for Ba that shows slightly higher values in the rift basalts. However, note that the Megezez basalts are much less primitive than the Gorfu-Mugher basalts. The rhyolites from the plateau have less extreme compositions as the acidic products from the rift. In general, all the silicic volcanics are alkaline to peralkaline.

Variation of trace elements vs. SiO_2 (Fig. 3) indicates a general increase in incompatible elements from basalts to rhyolites. Scattered values are observed in the intermediate and in the acid rocks.

Sr isotopic compositions are available only for the volcanic rocks from Debre Zeit. $^{87}Sr/^{86}Sr$ ranges from 0.70372 to 0.70869, displaying a well defined tendency to increase from mafic to salic rocks [8]. Small but significant isotopic variations are also observed in the basalts.

DISCUSSION

The main problems regarding the rocks under consideration are:

1. The genetic relationships between basalts and associated rhyolites for both the plateau and rift volcanism.

2. The genesis of basaltic magmas and their possible evolution during uprise to the surface.

3. The role of the continental crust in the genesis and evolution of the magmatism.

Debre Zeit

Variation diagrams for some key compositional parameters, together with geochemical models, are reported in Figs. 4 to 7. Variation of Sr vs Zr (Fig. 4) indicates that batch melting of metamorphic rocks from the Ethiopian basement (authors' unpublished data) are unable to give compositions as those of the rhyolites. The very high concentrations of Zr in the rhyolites can be obtained only if one assumes $D_{Zr} < 1$, and a very low degree of partial melting (< 10%). These assumptions are independent of the choice of the starting crustal rock. However, because of the high viscosity of acid magmas, it is unlikely that melts formed by less than 10% of crustal anatexis can separate from the residue [4, 22]. It is also unlikely that Zr behaves as an incompatible in crustal environment, since it is hosted in the lattice of several main and accessory crustal rock phases such as amphibole and zircon [e.g. 1].

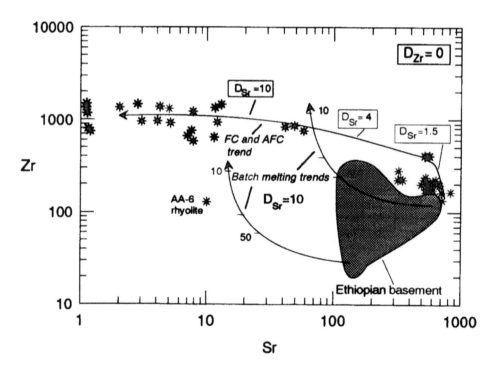

Figure 4. Sr vs Zr diagram for the Debre Zeit volcanics. The lines on the plot represent batch melting of the Ethiopian basement rocks, and magma evolution models. Fractional Crystallization (FC) and Assimilation plus Fractional Crystallization (AFC) trends are coincident. D_{Zr} and D_{Sr} are the bulk partition coefficients used in magma evolution and batch melting modeling. The composition of the Ethiopian Precambrian basement is from unpublished data of the senior author.

Only one rhyolite (AA-6) has low Zr, which makes it a likely candidate to represent anatectic liquids. Moreover, many Debre Zeit rhyolites have extremely low abundances of Sr. These cannot be generated by crustal anatexis, also if one assumes a strongly compatible behaviour of Sr during melting ($D_{Sr} = 10$). The bulk of the rhyolitic rocks fit quite well trends of Fractional Crystallization (FC) and Assimilation plus Fractional Crystallization (AFC), if a very incompatible behaviour for Zr and variable partition coefficients for Sr are assumed. More specifically, a requirement of the model is that Sr must become increasingly compatible with ongoing fractionation. The incompatible behaviour of Zr during fractional crystallization of alkaline magmas is supported by experimental evidence [24] and by the very high enrichments of Zr in the rhyolites. Increasing partition coefficients for Sr are plausible, because of the increasing role of feldspar fractionation from mafic to intermediate and acid liquids.

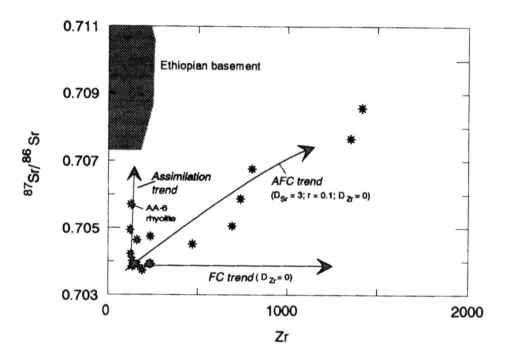

Figure 5. Zr vs. $^{87}Sr/^{86}Sr$ diagram for the Debre Zeit volcanics. The lines indicate calculated crustal assimilation, AFC, and FC trends. "r" is the ratio between amount of assimilated vs. crystallized material during AFC. D_{Zr} and D_{Sr} are the bulk partition coefficients used in the models.

Zr vs. $^{87}Sr/^{86}Sr$ diagram (Fig. 5) suggests that derivation of acid rocks from mafic ones is the products of AFC rather than simple fractional crystallization. Quantitative modeling, however, suggests that assimilation of crustal material was very small (mass ratio of assimilated vs crystallized: $r = 0.1$) and the AFC process was dominated by fractional crystallization. It must be also noted that the Debre Zeit rocks define two trends on the Zr vs. $^{87}Sr/^{86}Sr$ diagram: one connecting the basalts with rhyolites and one joining the basalts with the crustal rocks. This latter trend suggests an interaction between basalts and crustal

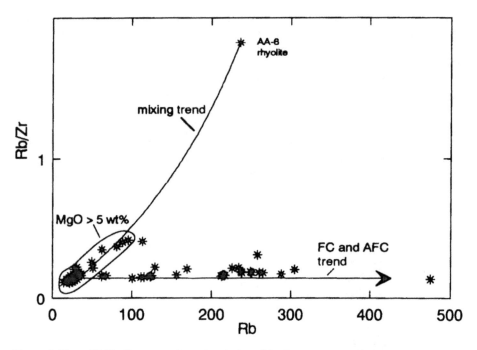

Figure 6. Rb vs. Rb/Zr diagram and geochemical models for the Debre Zeit volcanics.

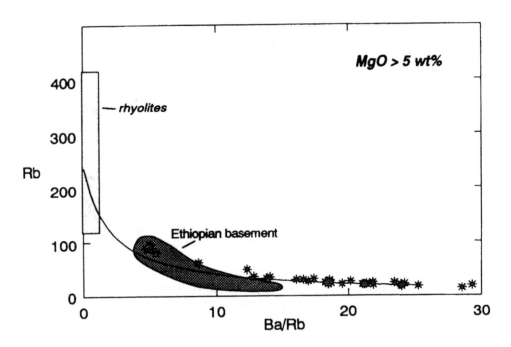

Figure 7. Ba/Rb vs. Rb diagram for the mafic volcanics (MgO > 5 wt%) from Debre Zeit. The line indicates a calculated mixing trend between basalts and the average rhyolite.

end-members. Rhyolite AA-6 plots along the second trend, which supports a genesis by crustal anatexis for this sample.

Fig. 6 shows Rb vs. Rb/Zr relationships for the Debre Zeit rocks. Two trends can be distinguished also on this diagram. The bulk of the basaltic rocks (MgO > 5%) plots along a mixing trend between the most mafic magmas and the rhyolite AA-6. This suggests that the chemical composition of the basalts was modified by mixing with rhyolitic liquids with a composition as AA-6, which most probably derived by melting of continental crust. The other samples, including intermediate and acid rocks, fit a fractional crystallization (or AFC) trend.

Ba/Rb vs. Rb diagram (Fig. 7) shows that the Debre Zeit basalts (MgO > 5%) display large variation of Ba/Rb ratios. This variation cannot be determined by fractional crystallization because both Rb and Ba behave as incompatible elements in the mafic systems. Neither it can be hypothesized that variable Ba/Rb is related to crustal assimilation, since the Ethiopian basement rocks have similar Ba/Rb ratios as the basalts. Accordingly, the most likely hypothesis is that this variation is related to mixing processes between mafic magmas and rhyolitic liquids. Fig. 7 shows that basalt compositions perfectly fit calculated mixing trend between mafic and acid magmas.

In summary, the petrogenetic picture that comes out from quantitative modeling of the data is as follows:

1. The bulk of the rhyolitic rocks was likely derived from basalts by AFC.

2. The amount of crustal material involved in the AFC was very small and the process was dominated by fractional crystallization

3. Based on only one sample, some acid magmas may derive by anatexis of crustal rocks.

4. The basaltic magmas have undergone significant modification of their geochemical characteristics due to mixing and interaction with crustal wall rocks.

5. Most probably, interaction between basaltic magmas and crustal material was not due to bulk crust assimilation but to mixing between basalts and crustal melts.

Mugher-Gorfu and Megezez
Isotopic studies on these rocks are still in progress. Accordingly, discussion will be only based on trace elements.

Variation diagrams with geochemical models are reported in Fig. 8. Sr vs. Zr and Y diagrams show that rhyolites have very high values of Y and Zr and low concentrations of Sr. As previously discussed (see Fig. 4), the high concentrations of Zr can be obtained by crustal melting only if very low degrees of partial melting are assumed, and if Zr behaves as an incompatible element during anatexis. The same holds true for Y. However, as mentioned above, it is unlikely that Zr is incompatible during partial melting. This worths also for Y, an element that is also hosted by several major and accessory crustal rock

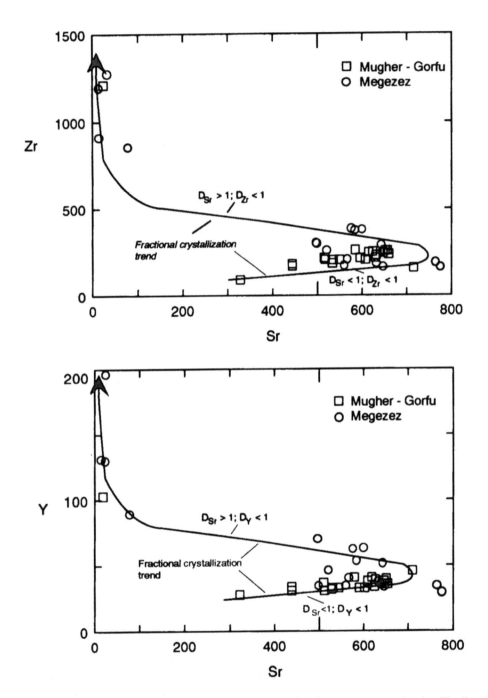

Figure 8. Sr vs. Zr and Sr vs. Y diagrams for the Mugher-Gorfu and Megezez volcanics. The lines are semiquantitative fractional crystallization trends. Variable D_{Sr} has been assumed during fractionation, in order to account for the variable role of plagioclase as separating phase. D_{Zr} has been assumed to be invariably lower than unity.

minerals such as zircon, allanite, amphybole and pyroxene [e.g. 1]. Accordingly, the most likely hypothesis is that rhyolites derived from basalts by fractional crystallization processes. The present data do dot allow to exclude some interaction between magma and crustal rocks, even though isotopic investigations are necessary in order to constrain this process.

Geochemical models indicate that fractional crystallization is initially dominated by the separation of mafic minerals (olivine and clinopyroxene). Accordingly, Sr, Zr and Y all are incompatible (D_{si} < 1) and there is a positive correlation of Sr vs Zr and Y in the mafic magmas. Successively, Sr becomes a compatible element because of the heavy fractionation of feldspars, and liquids are driven to Y- and Zr-rich and Sr-poor acid compositions.

Ba/Rb vs. Rb diagram (Fig. 9) shows large variations of Ba/Rb for mafic rocks, as observed in the Debre Zeit basalts. Again, this variation cannot be explained by either fractional crystallization or assimilation. Accordingly, mixing between mafic and acic magmas must be invoked.

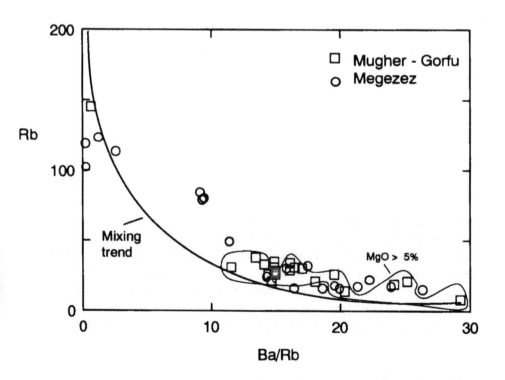

Figure 9. Ba/Rb vs. Rb diagram for the Mugher-Gorfu and Megezez volcanics. The line indicates a calculated mixing trend between the most mafic and the most acid compositions. The closed field indicates mafic rocks with MgO > 5 wt%.

In summary, a derivation of rhyolites from basalts by fractional crystallization (or AFC) seems a likely process also for the plateau and shield volcanism. Moreover, interaction between basalts and cogenetic rhyolites was a significant process in modifying the composition of basalts. Interaction between basalts and continental crust cannot be constrained by the present data. Isotopic investigations are in progress in order to verify this hypothesis.

CONCLUSIONS

The data discussed in this paper allow to put some constraints on the genesis and evolution of magmatism in Ethiopia. Variation of trace elements and Sr isotopic ratios indicate that the rift valley rhyolites are likely derived from associated basalts by AFC. Compositions of basalts have been modified by assimilation and mixing with rhyolitic rocks. These processes have modified Sr isotopic ratios and some incompatible element ratios such as Ba/Rb, Zr/Rb, etc. The same processes seem to have occurred during the earlier stages of volcanic activity, when the plateau and the shield volcanoes of the Termaber formation were built up. Partial melting of the continental crust appears to have played a minor role during the genesis of Debre Zeit rhyolites.

This petrogenetic model, if confirmed by further research, may have important geological implications. It is well known that ratios of incompatible elements such as Ba/Rb are currently used to constrain the source of continental flood basalts [e.g. 23, and references therein]. It is obvious that, if elemental ratios can be modified by low-pressure evolution, their use to infer source composition is severely restricted. In any case, the present data indicate that basalts can undergo important low-pressure modifications, even though they preserve their mafic composition. This stresses the need for detailed studies of evolutionary processes before making inferences on the composition of magma sources.

The derivation of rhyolites from basalts and the mixing between acid and mafic melts arise the problem related to the physical environments in which this process occurs. The large volumes of rhyolites, especially in the rift valley, require the presence of large magma chambers inside the crust. The same should be true for the plateau volcanism. The presence of large volcanoes with summit calderas along the Ethiopian rift valley (e.g. Gedemsa, Fantale, Gariboldi, etc.) supports this hypothesis. The problem is more complex for the plateau volcanics, which were emplaced by fissure eruptions. In any case, the occurrence of large crustal reservoirs would generate gravity and magnetic anomalies. The identification of such anomalies could be an objective for future investigations.

Acknowledgments

The authors thank the Technical Unit of the Italian Cooperation at the Italian Embassy in Addis Ababa for providing facilities during field work. The research has been financially supported by MURST, Project on Magma Genesis and Evolution in Regions of Continental Rift (A. Peccerillo).

REFERENCES

1. J.G. Arth. Behaviour of trace elements during magmatic processes. A summary of theoretical models and their applications. *J. Res. US Geol. Survey*, **4**, 41-47 (1976).

2. F. Barberi, L. Civetta, and J. Varet. Magmatic evolution in Afar. *Geodynamic evolution of the Afro-Arabic Rift System*, Atti Convegni Lincei, **47**, 455-462 (1980).

3. S.M. Berhe, B. Desta, C. Nicoletti, and M. Teferra. Geology, geochronology and geodynamic implications of the Cenozoic magmatic province in W and SE Ethiopia. *J. Geol. Soc. London*, **144**, 213-226 (1987).

4. J.D. Clemens, and C.K. Mower. Granite magma by transport propagation. *Tectonophysics*, **204**, 339-360 (1992).

5. C.V. Devey, and K.G. Cox. Relationships between crustal contamination and crystallization in continental flood basalt magmas with special reference to the Deccan Traps of Western Ghats, India. *Earth Planet. Sci. Letters*, **84**, 59-68 (1987).

6. M. Di Paola. The Ethiopian Rift Valley (between 7°.00' and 8°40' lat. North). *Bull. Volcanol.*, **36**, 317-356 (1972).

7. M. Franzini, L. Leoni, and M. Saitta. A simple method to valutate the matrix effect in X-Ray fluorescence analyses. *X-Ray Spectrometry*, **1**, 151-154 (1972).

8. M. Gasparon, F. Innocenti, P. Manetti, A. Peccerillo, and A. Tsegaye. Genesis of the Pliocene to Recent bimodal mafic-felsic volcanism in the Debre Zeit area, Central Ethiopia: volcanological and geochemical constraints. *African J. Earth Sci.*, **17**, 145-165 (1993).

9. I.L. Gibson. A review of the Geology, Petrology and Geochemistry of the Volcano Fantale. *Bull. Volcanol.*, **38**, 791-802 (1974).

10. W.K. Hart, G. WoldeGabriel, R.C. Walter, and S.A. Mertzman. Basaltic volcanism in Ethiopia: Constraints on continental rifting and mantle interaction. *J. Geophys. Res.*, **94**, 7731-7748 (1989).

11. C. Hofman, G. Feraud, R. Pik, G. Yirgu, D. Ayalew, C. Coulon, C. Deniel, and V. Courtillot. $^{40}Ar/^{39}Ar$ dating of Ethiopian traps. *Terra Abstracts*, **7**, 159 (1995).

12. H.E. Huppert, and R.S.J. Sparks. Cooling and contamination of mafic and ultramafic magmasduring ascent through continental crust. *Earth Planet. Sci. Letters*, **74**, 371-386 (1985).

13. T.N. Irvine, and W.R.A. Baragar. A guide to the elemental classification of common volcanic rocks. *Can. J. Earth Sci.*, **8**, 523-548 (1971).

14. A.B. Kampunzu, and P. Mohr. Magmatic Evolution and Petrogenesis in the East African Rift System. In: *Magmatism in Extensional Structural Settings*. A. B. Kampunzu and R. T. Lubala (Eds.). pp. 85-136, Springer, Berlin, (1991).

15. M.J. Le Bas R.W., Le Maitre A., Streckeisen A., and B. Zanettin. A chemical classification of volcanic rocks based on the total alkali-silica diagram. *J. Petrol.*, **27**, 745-750 (1986).

16. M.A. Menzies. Archean, Proterozoic and Phanerozoic lithospheres. In: *The Subcontinental Mantle*. M.A.Menzies (Ed.). pp. 11-125, Oxford University Press, Oxford (1990).

17. G. Merla, E. Abbate, A. Azzaroli, P. Bruni, P. Canuti, M. Fazzuoli, M. Sagri, and P.Tacconi.. *A geological map of Ethiopia and Somalia (1979) and comment with a map of major landforms*. C. N. R. Italy, Centro Stampa, Firenze, (1979).

18. P. Mohr, and B. Zanettin. The Ethiopian flood basalt province. In: *Continental flood basalts*. J.D. Macdougall (Ed.). pp. 63-110, Kluwer Acad. Publ., Dordrecht (1988).

19. A. Peccerillo, Y. Gezahegn, and A. Dereje. Genesis of acid volcanics along the Main Ethiopian Rift: a case history of the Gedemsa volcano. *Ethiop. J. Sci.*, **18**, 23-50 (1995a).

20. A. Peccerillo, M. Bekele, A. Tripodo, and E. Zimbalatti. Petrology and geochemistry of the Oligo-Miocene volcanism in the Mugher-Gorfu area, Central Ethiopia. *Ethiop. J. Sci.*, **18**, 79-101 (1995b).

21. E.M. Piccirillo, E. Justin-Visentin, B. Zanettin, J.L. Joron, and M. Treuil. Geodynamic evolution from plateau to rift: major and trace element geochemistry of the central-eastern Ethiopian Plateau volcanics. *Neus Jb.Palaont.Abh* , **158**, 139-179 (1979).

22. K.P. Skjerlie, and D. Johnston. Fluid absent melting behaviour of F-rich tonalitic gneisses. *J. Petrol.*, **34**, 785-803 (1993).

23. S. Turner, C. Hawkesworth, K. Gallagher, K. Stewart, D. Peate, and M. Mantovani. Mantle plumes, flood basalts, and thermal models for melt generation beneath continents: Assessment of a conductive heating model and applications to the Paranà. *J. Geophys. Res.*, **101**, 11503-11518 (1996).

24. E.B. Watson, and F.J. Ryerson. Partitioning of zirconium between clinopyroxene and magmatic liquids of intermediate compositions. Geochim. Cosmochim. Acta, **50**, 2523-2526 (1986).

25. R. White, and D. Mckenzie. Magmatism at rift zones: The generation of volcanic continental margins and flood basalts. *J. Geophys. Res.*, **94**, 7685-7729 (1989).

26. B. Zanettin, E. Justin-Visentin, M. Nicoletti, and E.M. Piccirillo. Correlation among Ethiopian volcanic formations with special references to the chronological and stratigraphical problems of the Trap Series. In: *Geodynamic Evolution of the Afro-Arabian Rift System*. Atti Convegni Lincei, **47**, 231 -252 (1979).

Proc. 30ᵗʰ Int'l Geol. Congr., Part 15   pp. 153–167
Li *et al.* (Eds)
© VSP 1997

Late Cenozoic Reactivation of the Early Pre-Cambrian Aldan Shield: Trace Element Constraints on Magmatic Sources beneath the Udokan Ridge, Siberia, Russia

SERGEI RASSKAZOV[1], ALEXEI IVANOV[1], ARIEL BOVEN[2], LUC ANDRÉ[3]

[1] *Institute of the Earth's Crust, Lermontov st. 128, Irkutsk 664033, Russia*
[2] *Vrije Universiteit Brussel, Pleinlaan 2, B-1050, Brussels, Belgium*
[3] *Musée Royal de L'Afrique Centrale, Steenweg op Leuven, 13 B-3080, Tervuren, Belgium*

Abstract

Late Cenozoic volcanic activity in the Udokan ridge is related to reactivation of the Kodar-Udokan and Chukchudu weak zones. The former locates in the area of the Baikal Rift System (BRS) overlapping the Proterozoic foredeep in the Aldan shield. The latter joins the Stanovoy suture which belongs to the Olekma-Stanovaya Orogenic System (OSOS) reactivated due to collision of the Bonin and Honshu arcs. Diverse volcanic pulses in the weak zones indicate independent tectonic activity in the BRS and OSOS. Trace element abundances in lavas from both zones have been studied by ICP-MS technique. The initial 14 Ma olivine melaleucetite magmas in the Kodar-Udokan zone were inferred to represent phlogopite-bearing mantle source depleted in Pb and Y (Ce/Pb = 20; Y/Ho = 22). Alkali olivine basalts and basanites erupted between 3.2 and 2.4 Ma exhibited hybrids of two or three end-members including, beside the Pb-Y depleted component, those of the lower crust and similar to the primitive mantle. In the Chukchudu zone, the 4.0-3.5 Ma basanites were dominated by the OIB-like component. During the last 1.8 Ma, the low-Y material melted. It is suggested that lavas of the Kodar-Udokan zone represent the continental mantle which underlain the Aldan shield in the Archean, but partly or completely modified by the Proterozoic processes of deep tectono-thermal reactivation. Lavas of the Chukchudu zone exhibit lithospheric mantle material formed during deep Phanerozoic convection and/or mantle material of the Baikal-Vitim terrane subducted beneath the Archean crust of the Aldan shield in the Paleozoic.

Keywords: Baikal Rift System, Cenozoic, alkaline volcanic rocks, trace elements

INTRODUCTION

The chemically distinct highly refractory lithosphere keel beneath the stable Archean cratons is argued to be produced early in the Earth's history. In contrast, tectonically active circumcratonic Proterozoic and Phanerozoic terranes are usually characterised by less refractory mantle. Geometry of lithospheric domains beneath the Archean cratons and adjacent areas has been recognised through study of chemical and isotopic characteristics of the Late Cenozoic basalts, basalt-born xenoliths and kimberlite-born xenoliths in North America, Southern Africa and Eastern China [11, 12, 26].

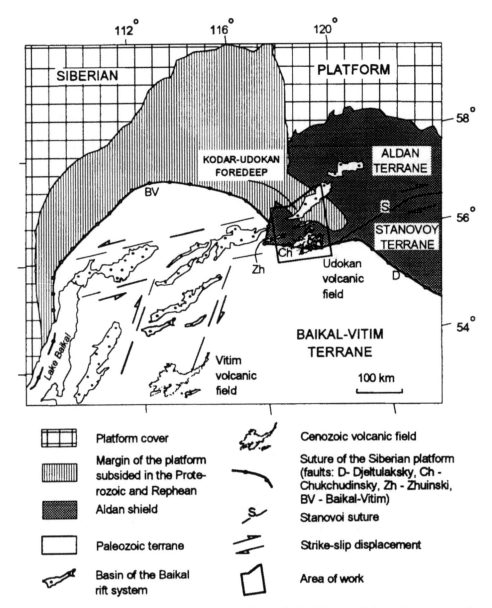

Figure 1. Structural setting of the Udokan volcanic field. Scheme of the major structural elements modified and simplified after Melnikov et al. [10], Gusev and Hain [1].

In Southern Siberia, Cenozoic tectonic reactivation strongly affected the southern margin of the Siberian platform and adjacent folded terranes. Concentration of the extension at the south-eastern suture of the Siberian platform resulted in the formation of the deepest depression of Lake Baikal. It is located in the central part of the Baikal Rift System (BRS) [9]. The south-western and north-eastern parts of the BRS are expressed with more shallow basins accommodated within the Phanerozoic terranes. The north-eastern terminus of the rift structures penetrates into the western Aldan shield (Fig. 1). The rest part of this shield is affected by compression and exhibits the Olekma-Stanovaya Orogenic System (OSOS) [5, 21].

Volcanic fields of the BRS occupy mostly Phanerozoic terranes. One may expect to identify the Archean mantle and/or crustal components only in the Cenozoic volcanic rocks from the Aldan shield. To assess the nature of mantle and crust beneath this area, in this paper we present trace element data on lavas erupted in the Udokan volcanic field (UVF).

This volcanic field occupies an area of about 3, 000 km² on the Udokan ridge southeast of the Chara basin (Fig. 2). In the Middle Miocene through Pliocene, lavas have formed a shield as thick as 700 m. Volume of the erupted volcanic material exceeds 430 km³. The lavas buried 500 m deep river paleovalleys. An uplift and erosion at about 2-1 Ma converted the volcanic stratum into high plateau. The lava sequences are well exposed in the present-day river canyons. During the Quaternary, minor volcanic activity rejuvenated in the western margin of the volcanic field.

Structural setting of the UVF
The southern suture of the Siberian platform is traced along the Djeltulaksky, Stanovoi, Chukchudu, Zhuinsky and Baikal-Vitim faults (Fig. 1). The latter bounds the subsided platform margin overlain by the Proterozoic and Rephean formations. Its uplifted part is represented by the Aldan shield. The local subsidence in the western edge of the shield is expressed with the Kodar-Udokan foredeep. The Baikal-Vitim terrain has been accreted to the Siberian Platform in the Paleozoic.

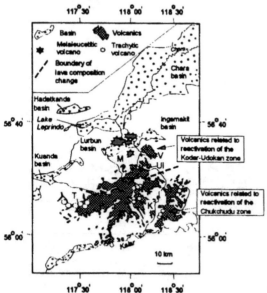

The Aldan shield is subdivided into the Aldan and Stanovoi terranes separated from each other by the Stanovoi suture. In the late Cenozoic, oblique (NW-SE) compression in the OSOS resulted in a reactivation of this suture as a right-lateral strike-slip fault [5]. Simultaneously, the north-eastern part of the BRS underwent the NW-SE extension and left-lateral strike-slip faulting [22].

The NW-SE extension in the BRS was believed to be triggered by the Indo-Asian collision and enforced by an upwelling of the hot mantle material [9, 17]. The NW-SE compression in the OSOS seemed to be created by collision between Bonin and Honshu arcs. This collision took place at 12-10 Ma. Its mid-plate effect may explain the drastic

Figure 2. Structural position of the Udokan volcanic field and location of the alkaline ultramafic and trachytic volcanoes. Location of the studied crossections: M, Munduzhak; UI, Upper Ingamakit; V, Vakat; K, Kanksa; T, Turuktak.

Late Miocene change of stress field in the Japan Sea, East China and Far East of Russia, as well as a co-existence of two stress fields in the eastern Eurasian plate and opposite slip motions in the north-eastern BRS and in the OSOS [6, 21, 24].

The northern part of the UVF overlaps the Kodar-Udokan weak zone. The rest part of the volcanic field is in the Chukchudu weak zone. Spatial locations of the Chara basin and smaller depressions of the north-eastern BRS are inferred to be strictly controlled by the reactivated Kodar-Udokan weak zone. Therefore, volcanism of this area is also to be related to rifting [18]. On the other hand, reactivation of the Chukchudu weak zone apparently reflects a dynamics of the OSOS. Here, a volcanic activity might be dominated by the episodic collision-derived reactivation of the Stanovoi suture [18, 20].

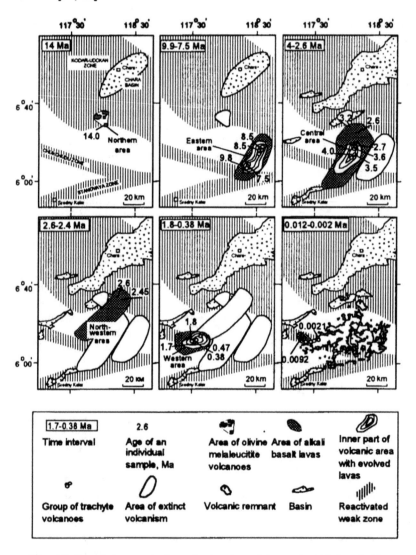

Figure 3. Space-time evolution of volcanic activity in the Udokan field.

Timing of volcanic activity
The results of K-Ar dating [18, 20] show an asynchronous volcanic activity in the reactivated Kodar-Udokan and Chukchudu zones, probably due to different timing of tectonic reactivation in the BRS and OSOS (Fig. 3). The first eruptions in the Kodar-Udokan zone took place at about 14 Ma simultaneously with initiated subsidence of the Chara basin. After long period of volcanic quiescence, eruptions rejuvenated again between 3.2 and 2.4 Ma. In the Chukchudu zone, the volcanism started about 10 Ma. There was a stepwise shift of volcanism along the Chukchudu zone from the eastern part of the field to the western one. The time intervals of volcanic activity within three different areas are at 9.9-7.5 Ma (the eastern part of the UVF), 4.0-2.6 Ma (the central part) and from 1.8 Ma to the Holocene (the western part). According to ^{14}C dating, the latest volcanic eruptions took place in the UVF between 12,000 and 2100 B.P.

LAVA SEQUENCES

Volcanic units of the Kodar-Udokan zone
This area was enveloped with mafic and ultramafic volcanism. Compositionally, lavas range from olivine melaleucetite through basanite, alkali olivine basalt and hawaiite to olivine tholeiite. No evolved lavas have been found. *The oldest unit* (14 Ma) is represented with nine volcanoes scattered in area of 120 km^2 on the topographic elevation between the Ingamakit and Lurbun depressions (Fig. 2). Some of the cones are as high as 120 m. Three other units belong to the lava stratum which filled the Early Pliocene erosional paleovalleys. In the lower parts of the "paleovalley" volcanic stratum, lava flows are interbedded with sedimentary lenses. The outcrops of the Munduzhak, Upper Ingamakit and Vakat sequences number from 20 to 40 basalt layers. Most of the flows are 2-4 m thick. Some of them exceed 30 m. A location of the studied sequences is shown on Fig. 2. A thickness of *the lower unit* (3.2-2.7 Ma) varies from 8 to 100 m and more. *The middle unit* (~2.6 Ma) is 20-30 m thick. One can identify the unit due to coarse-grained doleritic structure of its rocks which contain large (up to 1 cm) plagioclase. *The upper unit* (2.6-2.4 Ma) comprises an uniform lava package and "summit" cinder and scoria cones. The lava package exceeds 200 m. The largest cone is 200 m high.

Volcanic units of the Chukchudu zone
In the Chukchudu zone, two differentiated series have been distinguished [16]. The moderately alkaline series comprises a continuous range of compositions from the alkali olivine basalt through hawaiite, mudgearite and benmoreite to trachyte. The highly alkaline one is represented with basanite and phono-basanite. Some basanites are of ultramafic compositions and show Mg# as high as 70. The evolved lavas erupted during all three volcanic episodes. Trachytic volcanoes form loci oriented northeast-southwest in the eastern UVF, north-south in its central part and west-east in the western part (Fig. 2).

In this zone, we focus our study on the lava sequences of the central and western parts of the UVF. *The Kanksa unit* (4.0-3.5 Ma) of the central UVF is a 600 m thick stratum exposed on the eastern side of the Kanksa river. Mafic and ultramafic lavas are interbedded with pyroclastic material. Trachyte flows occur both at the foot and on the top of the sequence. *The Turuktak unit* (~1.8 Ma) represents a range of mafic members of the moderately alkaline differentiated series (alkali olivine basalts and hawaiites).

CHEMISTRY OF VOLCANIC ROCKS

Analytical techniques, previous and new geochemical data
Our previous study [18] was focused on the temporal variations of the lava compositions in the northern UVF (Kodar-Udokan zone). Thirty four samples from the Munduzhak and Upper Ingamakit volcanic sequences had been selected to cover the compositional ranges of all four volcanic units. Major elements, Ni, Co and Cr were analysed by XRF method in the Royal Museum of Central Africa, Tervuren and in the Institute of Geochemistry, Irkutsk. Concentrations of the REE, Ba, Rb, Pb, Nb, Ta, Zr, Hf, V, W, Th and U were measured by ICP-MS technique in the Royal Museum of Central Africa. Sr abundances were determined by XRF method in the Institute of the Earth's Crust and Institute of Geochemistry, Irkutsk.

For this study, 10 additional samples from the Vakat sequence of the Kodar-Udokan zone have been analysed. Geochemistry of the evolved lava verities from the volcanic units of the Chukchudu zone is to be considered in another paper. The purpose of this one is to report data only on the ultramafic and mafic lavas which occur both in the Kodar-Udokan and Chukchudu zones. The Kanksa and Turuktak units are represented with 18 and 8 samples, respectively. All additional samples have been analysed by the same techniques as in the previous work. Representative analyses of the units are reported in Table 1.

Classification and terminology of volcanic rocks
We follow a common definition of volcanic rocks based on modal mineralogy, normative compositions [16] and total alkali - silica diagram (TAS diagram) [7, 8]. In the TAS diagram (Fig. 4), analyses have been recalculated to 100% on water-free basis and plotted as different age groups of rocks from the Kodar-Udokan and Chukchudu zones. Normative compositions and *mg*-numbers (Mg#=atomic ratio $100Mg/(Mg+Fe^{2+})$) were calculated assuming $Fe^{3+} = 0.15Fe_{total}$.

Olivine melaleucitite is a strongly undersaturated alkaline ultramafic lava with normative and modal leucite. In thin sections, this rock is feldspar-free or shows tiny sanidine and/or plagioclase laths. Up to 5% of normative $Ab = 100an/(an+ab)$ may be calculated [7]. *Basanite* is a feldspar-bearing highly alkaline lava with normative *ne* of 5-17%. *Alkali olivine basalt* and *hawaiite* possess up to 5% of *ne*. In the former, $An = 100an/(an+ab)$ is more than 50%. In the latter, *An* varies from 50 to 30%. In the TAS diagram, these rocks are plotted on both sides from the formal IUGS boundary between trachybasalt and basanite (Fig. 4). *Olivine tholeiite* is an *ol-hy*-normative basaltic lava. In the TAS diagram, these rocks are plotted just above the McDonald and Katzura line separating alkaline and tholeiitic volcanic rocks of the Hawaiian islands.

Table 1. Representative analyses of the UVF rocks

Sample Rock Age, Ma	Ud7/18 OML	P258/2 OT 3.2-2.7	P240/2 OT 2.6	U-94-3 BS 2.6-2.4	P335/16a BS	P335/22 OT 4-3.5	Ud3/4 AOB <1.8
SiO$_2$	41.99	47.17	48.66	45.67	42.80	47.60	47.52
TiO$_2$	2.75	2.00	2.18	2.73	2.42	2.00	2.33
Al$_2$O$_3$	11.90	15.55	16.14	16.60	13.76	15.12	15.68
Fe$_2$O$_3$	3.44	2.76	1.98	4.70	6.86	4.23	6.13
FeO	8.08	8.46	9.41	7.07	6.16	6.69	6.55
MnO	0.18	0.15	0.15	0.15	0.20	0.16	0.17
MgO	12.90	7.96	7.00	6.27	9.16	8.13	7.67
CaO	11.20	8.50	9.17	7.24	11.06	8.61	8.61
Na$_2$O	3.22	2.56	3.15	4.88	4.06	3.09	3.94
K$_2$O	2.26	1.45	0.99	2.52	1.79	1.43	1.68
P$_2$O$_5$	0.79	0.28	0.24	0.84	0.84	0.40	0.47
LOI	1.30	1.90	0.30	0.86	2.11	3.05	0.02
Total	100.01	98.74	99.37	99.53	101.22	100.51	99.09
Sr	1080	467	408	893	n.d.	n.d.	810
Ba	838	434	248	608	719	407	517
Rb	43	18.5	12.4	37	53.2	18.4	26
Pb	7.4	5.3	3.76	10	5.23	5.24	7.9
Nb	57.8	28.4	19.4	68	82.1	27.29	32.7
Ta	4.03	1.85	1.11	4.5	4.61	1.77	2.52
Zr	155	139	119	322	241	136	169
Hf	4.89	3.98	3.27	6.3	5.62	3.91	5.46
Ni	292	122	82	n.d.	118	131	144
Co	79	71.3	67.1	n.d.	44.8	46	64
Cr	410	199	155	n.d.	n.d.	201	155
V	178	181	192	177	153	186	137
La	53	27	15.9	48	70	24.2	37
Ce	101	56	36	91	141	53.5	68
Pr	11.8	6.9	4.4	10.5	14.9	6.19	8.2
Nd	46	29	19.5	40	56.9	25.9	34
Eu	2.32	2.1	1.71	2.4	3.2	1.92	2.04
Sm	8	6.3	4.95	7.5	9.68	5.33	6.7
Gd	6.2	6	5.07	6.2	9.6	5.76	5.8
Dy	4.42	5.35	4.67	4.8	5.98	4.21	4.93
Ho	0.77	0.88	0.77	0.84	1.02	0.76	0.92
Er	1.92	2.29	2	2.1	2.51	1.99	2.46
Yb	1.44	1.83	1.62	1.63	1.89	1.68	2.09
Lu	0.19	0.24	0.22	0.22	0.24	0.21	0.28
W	2.34	1.31	1.61	1.08	1.96	n.d.	2.48
Th	7.1	3.12	1.31	5.5	8.49	2.7	4.65
U	2.05	1.41	0.66	1.55	2.61	0.91	1.45
Y	17.1	19	20	24	27.4	19.3	19.9

OML - olivine melaleucitite, OT - olivine tholeiite, BS - basanite, AOB - alkali olivine basalt. n.d. - not determined.

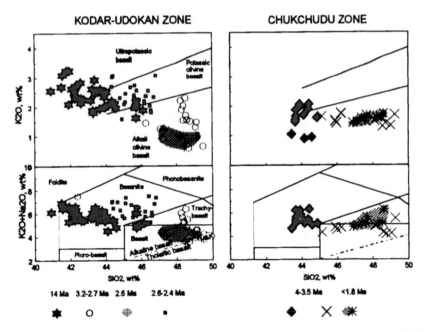

Figure 4. Temporal variations of K_2O, K_2O+Na_2O vs SiO_2 in the lavas from the Udokan volcanic field. Separating lines in the K_2O vs SiO_2 plot after Moore and Dodge [14]. Separating lines in the TAS diagram after Le Bas and Streckeisen [7]. M&K - McDonald and Katzura line separating tholeitic and alkaline series of the Hawaiian islands.

Major elements
The 14 Ma rocks of the Kodar-Udokan zone have relatively high Mg# (65-71), lower silica concentrations and vary in composition from olivine melaleucitite to basanite. In the 3.2-2.7 Ma unit, alkali olivine basalts predominate. Package of basanites with 5-8% *ne* is found in the upper part of the unit. Hawaiites and olivine tholeiites are rare. The rocks of the 2.6 Ma unit are olivine tholeiites and alkali olivine basalts with higher silica contents. This stratigraphic level involves also a single olivine leucitite flow. The 2.6-2.4 Ma unit varies from basanite to olivine leucitite compositions. In the Chukchudu zone, the mafic and ultramafic lavas of the Kanksa unit are mostly basanites. The alkali olivine basalt flows and packages are recorded in six levels of the section. The Turuktak unit is represented with alkali olivine basalts, hawaiites and basanites with 5-8% of *ne*.

In K_2O vs SiO_2 plot (Fig. 4) highly alkaline lavas of the Kodar-Udokan zone occupy ultrapotassic and potassic fields. The lavas include sanidine megacrysts with orthoclase component up to 62%. In the moderately alkaline lavas of this area, the potassium content is lower. Their inclusions are megacrysts of plagioclase. In the Chukchudu zone, the lower potassic content is a characteristic of both moderately and highly alkaline lavas. The latter contain abundant anorthoclase and albite megacrysts with an orthoclase component as low as 8% [16].

Trace elements
The alkaline nature of the volcanic rocks assumes high abundances of trace elements which are incompatible during melting of mantle peridotite. Diagrams of Fig. 5-9

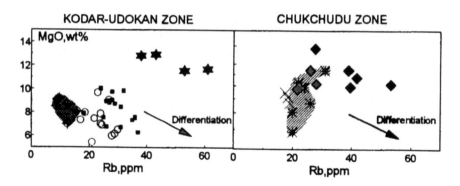

Figure 5. Temporal variations of MgO vs Rb in the UVF lavas. Symbols as in Fig. 4.

reveal effects of fractional crystallisation, varying degree of partial melting and heterogeneities in the source regions.

MgO vs Rb plot
During fractional crystallisation highly incompatible element concentrations in mafic magmas (Rb, U, Th etc.) usually increase as MgO content decreases. The MgO vs Rb plot (Fig. 5) shows that age groups of olivine melaleucitites, alkali olivine basalts and basanites in the Kodar-Udokan zone and basanites, alkali olivine basalts from the Chukchudu zone can not be related as parent and daughter liquids, although slight differentiation is responsible to the negative correlations between MgO and Rb within each group. Alkali olivine basalts from the Turuktak unit do not indicate differentiation and are comparable with the olivine melaleucitites to alkali olivine basalts trend of the lavas from the Kodar-Udokan zone.

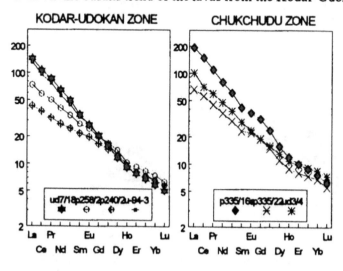

Figure 6. Chondrite-normalised diagram for selected samples. Normalising values after Sun and McDonough [23]. Symbols as in Fig. 4.

REE elements
The volcanic units are well defined on chondrite-normalised REE plot (Fig. 6). The oldest olivine melaleucitite from the Kodar-Udokan zone (sample ud7/18) displays the highest concentrations of the light REE (LREE) and the lowest ones of heavy REE (HREE). In the alkali olivine basalt (sample p258/2) and olivine tholeiite (sample p240/2) from the lower and middle "paleovalley"

units respectively, LREE decrease and HREE increase. As a result REE patterns for the alkali olivine basalt, olivine tholeiite and olivine melaleucitite intersect. The REE profile of the basanite (sample u-94-3) from the upper unit is plotted between those of the olivine melaleucitite and alkali olivine basalt. The olivine melaleucitite and basanite lines intersect. The chondrite-normalised patterns for the basanite (sample p335/16a) and alkali olivine basalt (sample p335/22) from the Kanksa unit do not intersect. The line of the alkali olivine basalt from the Turuktak unit (sample ud3/4) shows relative increase of HREE and intersects the basanite line. The crossing patterns with high LREE may indicate partial melting of garnet peridotite [2].

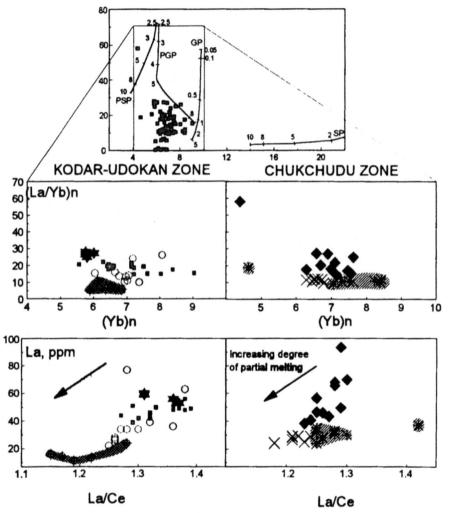

Figure 7. Temporal variations of $(La/Yb)n$ vs $(Yb)n$ and La vs La/Ce in the UVF lavas. Symbols as in Fig. 4. Insert shows partial melting curves for hypothetical mantle sources after Zhang et al. [27]. SP, spinel peridotite; GP, garnet peridotite; PSP, phlogopite spinel peridotite; PGP, phlogopite garnet peridotite.

Increasing degrees of melting in garnet and spinel peridotites are expressed with $(La/Yb)_n$ and $(Yb)_n$, respectively (Inset in Fig. 7). In $(La/Yb)_n$ vs $(Yb)_n$ plot, the UVF lavas occupy the field of relatively high extent of melting in garnet and spinel peridotites. High $(La/Yb)_n$ and low $(Yb)_n$ of OML may indicate contribution to the primary melts of liquids from a phlogopite-bearing peridotite. Large variation of $(La/Yb)_n$ in 3.2-2.6 Ma lavas reflects melting of rocks in the depth from garnet facies of the mantle (high $(La/Yb)_n$ and high $(Yb)_n$) to the lower crust (low $(La/Yb)_n$ and low $(Yb)_n$). More shallow liquids are characterised by relatively high degree of melting. It is displayed also with decrease of $(La/Ce)_n$ ratio as La decreases. Lavas of the next (2.6-2.4 Ma) magmatic episode reveal a trend to high values of $(Yb)_n$ probably due to contribution of the low degree liquids from spinel peridotite.

Compared to alkali olivine basalts, basanites of the Kanksa (4.0-3.5 Ma) unit have definitely high $(La/Yb)_n$, $(La/Ce)_n$, and La abundance. This may be interpreted as manifestation of varying degree of melting. High $(La/Yb)_n$ in some basanite samples may be related to contribution of low degree melts from garnet peridotite. Another basanite trend to the lower $(Yb)_n$ values perhaps results from melting of the phlogopite-bearing peridotite. During the last 1.8 Ma, regime of melting was different. The alkali olivine basalts of the Turuktak unit show relative increase of $(Yb)_n$ and $(La/Ce)_n$ due to contribution of the lower degree liquids from spinel peridotite.

U vs Nb, Pb vs Ce and Y vs Ho plots
U and Nb, Pb and Ce, Y and Ho have similar behaviours during mantle melting and crystal fractionation. These elements are indicative of variations of heterogeneities in source regions. Nb/U ratio of the oceanic basalts is 47±10 and in the crust decreases up to 10 [3, 15]. Ce/Pb ratio of the oceanic basalts is 25±5 decreasing in the lower crust up to 4 [3, 24]. The bulk silicate earth is characterised by Ce/Pb ratio of 10.7±0.5 [14]. Y/Ho ratio of the oceanic basalts and primitive mantle is about 27-28 [23]. In the lower crust, it decreases up to 25 [24].

In U-Nb and Pb-Ce diagrams (Fig. 8), lavas of the Kodar-Udokan zone are distributed between the compositions of the oceanic basalts and the lower crust. The olivine melaleucitites are plotted in the beginning of the highly and moderately alkaline trends. The former exhibits a relative increase of the Pb content and nearly constant Ce concentrations from this point to the Ce/Pb value of the primitive mantle. The latter displays a wider range of Pb concentrations as Ce content decreases. Ce and Pb abundances of the 2.6 Ma unit are close to the average lower crust by Taylor and McLennan [24]. Lavas from the Chukchudu zone display another trends. Both Nb/U and Ce/Pb ratios of the BS are partly in the range of the oceanic basalts. The alkali olivine basalts interlayered with the basanites in the Kanksa unit are displaced towards the lower crust (or primitive mantle). The younger alkali olivine basalts from the Turuktak unit indicate more prominent shift outwards the field of the oceanic basalts.

Y and Ho are more compatible elements. The olivine melaleucitites indicate anomalously low Y/Ho ratios (22). The others rocks from the Kodar-Udokan zone cover the range of Y/Ho ratios between the olivine melaleucitites and the oceanic basalts. Y and Ho abundances of the 2.6 unit are similar to the lower crust. Similar wide range of Y/Ho ratios is a characteristic of the rocks from the Kanksa unit. The

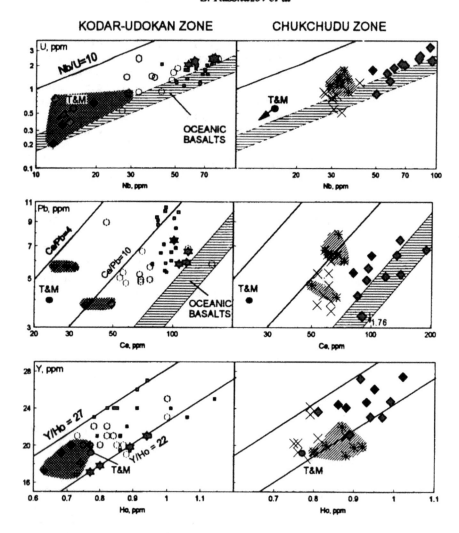

Figure 8. Temporal variations of U vs Nb, Pb vs Ce and Y vs Ho in the lavas from the Udokan volcanic field. T&M is the lower crust after Taylor and McLennan [24].

alkali olivine basalts from the Turuktak unit show the most significant shift from the oceanic basalt compositions and in this respect are comparable to the olivine melaleucitites.

DISCUSSION

Neodymium isotope compositions of the UVF lavas correlate with their chemistry [3, 18, 19]. Basanites are originated from the depleted mantle (εNd from +4.68 to +3.75) whereas alkali olivine basalts and olivine tholeiites contain components of the less depleted mantle and lower crust (εNd from +3.69 to -7.2). $^{87}Sr/^{86}Sr$ ratios in these rocks vary from 0.703645 to 0.704024. The lowest εNd values (-7.2) and chemical data of Fig.

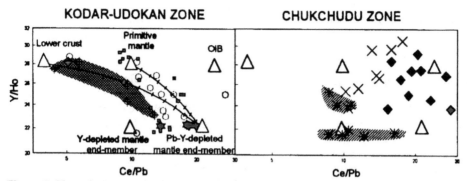

Figure 9. Hypothetical end-members of the UVF magmas in the Y/Ho vs Ce/Pb diagram. Symbols as in Fig. 4. OIB and primitive mantle after Sun and McDonough [23], lower crust after Taylor and McLennan [24]. See text for explanation of mixing lines.

4-8 on the 2.6 Ma olivine tholeiites from the Kodar-Udokan zone are indicative of lower crust contribution [18]. One basanite sample (p269/2a) from this zone shows the signature of the enriched mantle (εNd -3.52, $^{87}Sr/^{86}Sr$ 0.704349).

Lavas of the Kodar-Udokan and Chukchudu zones indicate notably diverse lead isotope compositions. The 2.6-2.4 basanites from the former have the most unradiogenic lead ($^{206}Pb/^{204}Pb$ 17.792-17.935, $^{207}Pb/^{204}Pb$ 15.430-15.491, $^{208}Pb/^{204}Pb$ 37.611-37.872). $^{207}Pb/^{204}Pb$ ratio in the 3.2-2.6 alkali olivine basalts and olivine tholeiites is lower and ranges from 15.3772 to 15.3800. In the rocks from the central and eastern parts of the Chukchudu zone, lead is more radiogenic [19]. In the UVF, there are no compositions with significantly negative εNd and low values of $^{206}Pb/^{204}Pb$ which would be a characteristic of the Archean continental lithosphere [12, 26].

The available geochemical and isotopic data reveal apparent diversity in evolution and composition of the late Cenozoic magmatism in the Kodar-Udokan and Chukchudu zones which may be referred to different lithospheric domains. The origin of the domains beneath the Udokan ridge is satisfactorily explained in context of the tectonic framework (Fig. 1). Lavas of the Kodar-Udokan zone represent the mantle material underlaying the Aldan shield in the Archean, but partly or completely modified by the Proterozoic and probably later processes of deep tectono-thermal reactivation. Lavas of the Chukchudu zone exhibit the lithospheric mantle material formed during deep convection during the Phanerozoic and/or the mantle material of the Baikal-Vitim terrane subducted beneath the Archean crust of the Aldan shield.

The lithosphere of the BRS is as thin as 40-50 km. It is suggested that beneath the axial rift basins, an upwelling of the partly molten asthenosphere reaches a foot of the crust [28]. This is consistent with geochemical data showing melting of the shallow mantle and lower crust. The lavas from both the Kodar-Udokan and Chukchudu zones are predominated by high degree partial melts from the mantle peridotites. Some average composition may be considered as a hypothetical primary liquid which has been contributed with lower degree partial melts from anhydrous garnet and spinel peridotites as well as phlogopite-bearing (amphibole-bearing) peridotites in the process of magmatic evolution (Fig. 7). Thus, lavas of the UVF are

S. Rasskazov et al

melt mixtures of two or more end-members. On the other hand, some volcanic episodes (for example the 14 Ma olivine melaleucitites and alkali olivine basalts younger than 1.8 Ma) show geochemically unique compositions reflecting a local batch melting in the mantle.

The possible end-members are shown in the Y/Ho vs Ce/Pb plot (Fig. 9). The 14 Ma olivine melaleucitites of the Kodar-Udokan zone were derived from the lithospheric mantle depleted in Pb and Y (Y/Ho = 22; Ce/Pb = 20). Relative decrease of Ce/Pb may indicate the Y-depleted component with Ce/Pb ratio about 10. The younger alkali olivine basalts and basanites are distributed along the mixing line between the Pb-Y-depleted component and one corresponding to the primitive mantle. The 2.6 Ma alkali olivine basalt-olivine tholeiites are at the mixing line between the same component and the lower crust. In the Chukchudu zone, the Pb-Y- and Y-depleted end-members are also available. They are well expressed in the youngest lavas of the western Chukchudu zone. In contrast to the Kodar-Udokan zone, the OIB-like component predominates in the basanites of the central Chukchudu zone. The alkali olivine basalts from the same area are shifted towards the composition of the primitive mantle (or lower crust).

The OIB-like magmas from the Chukchudu zone might represent a material of the sub-lithospheric hot plume. In this respect, a persistent shift of volcanism from the east to the west should be interpreted in connection with the eastward motion of the Eurasian plate above the fixed hot plume with the local velocity about 0.8 cm/yr [17].

CONCLUSIONS

Volcanic activity in the Udokan field was controlled by independent tectonic processes in the BRS and OSOS. Extension in the former was concentrated in the Kodar-Udokan weak zone where prominent tectonic and volcanic pulses took place at about 14 and 3.2-2.4 Ma. The collision derived tectonic pulses of the OSOS were exhibited in the Chukchudu weak zone with volcanic episodes of 10-7, 4-2.6 Ma and 1.8 Ma to the Present.

Diverse late Cenozoic magmatism in the Kodar-Udokan and Chukchudu zones may be due to its development in two different lithospheric domains. Lavas of the Kodar-Udokan zone represent the mantle material from the Archean Aldan shield partly or completely modified by the Proterozoic and probably later processes. Lavas of the Chukchudu zone exhibit the lithospheric mantle material formed during deep Phanerozoic convection and/or the mantle material of the Baikal-Vitim terrane subducted beneath the Archean crust of the Aldan shield in the Paleozoic.

ACKNOWLEDGEMENTS

This work was initiated as a part of a collaborative Russian-Belgian research supported by INTAS grant N 93-134. S. Rasskazov and A. Ivanov were funded by RFBR grants N 95-05-14277, 97-05-65332 and 97-05-65331. Some samples from the Munduzhak and Turuktak sequences were donated by V. Ivanov. We are most grateful for the help and support of N.A. Logatchev and J. Klerkx.

REFERENCES

1. G.S. Gusev and V.E. Hain. On relations between Baikal-Vitim, Aldan-Stanovoi and Mongol-Okhotsk terranes (South of Mid-Siberia), Geotectonica 5, 68-83 (1995). (in Russian)
2. G.N. Hanson. An approach to the trace element modelling using a simple igneous system as an example. In: Geochemistry and mineralogy of rare earth elements. B.R. Lipin and G.A. McKay (Eds.). Reviews in mineralogy 21, pp. 79-97 (1989).
3. N. Harris, D.S. Coleman, S.A. Bowring, S. Rasskazov. Geochemical evidence for Lithospheric contributions to volcanism at Udokan and Vitim fields, Baikal rift, Siberia (abs.), EOS, AGU Spring Meeting 87, S287 (1996).
4. A.W. Hofmann. Nb in Hawaiian magmas: constraints on source composition and evolution, Chemical Geology 57, 17-30 (1986).
5. V.S. Imaev, L.P. Imaeva, B.M. Kozmin and K. Fudzhita. Active foults and modern geodynamics of the Yakutian seismical belts, Geotectonika 2, 57-71 (1994). (in Russian).
6. L. Jolivet, K. Tamaki and M. Fournier. Japan Sea, opening history and mechanism: A synthesis, J. Geophys. Res. 99, 22,237-22,259 (1994).
7. M.J. Le Bas. Nephelinitic and basanitic rocks, J. Petrology 30, 1299-1312 (1989).
8. M.J. Le Bas and A.L. Streckeisen. The IUGS systematics of igneous rocks, J. Geol. Soc. London 148, 825-833 (1991).
9. N.A. Logatchev. History and geodynamics of the Lake Baikal rift in the context of the Eastern Siberia rift system: a reviw, Bull. Centres Rech. Explor.-Prod. Elf. Aquitaine 17, 353-370 (1993).
10. A.I. Melnikov, A.M. Mazukabzov, E.V. Sklyarov and E.P. Vasiljev. Baikal rift basement: structure and tectonic evolution, Bull. Centres Rech. Explor.-Prod. Elf. Aquitaine 18, 99-122 (1994).
11. M.A. Menzies. Cratonic, circumcratonic and oceanic mantle domainas beneath the estern United States, J. Geophys. Res. 94, 7899-7915 (1989).
12. M.A. Menzies. Archean, Proterozoic, and Phanerozoic lithospheres. In: Continental mantle. M.A. Menzies (Ed.). pp. 67-86. Clarendon press, Oxford (1990).
13. D.M. Miller, S.L. Goldstein and C.H. Langmuir. Cerium/lead and lead isotope ratios in arc magmas and the enrichment of lead in the continents, Nature 368, 514-519 (1994).
14. J.D. Moore and F.C.W. Dodge. Late Cenozoic volcanic rocks of the southern Sierra Nevada, California: I. Geology and Petrology: summary, J. Geol. Soc. Amer. Bull. 91, 515-523 (1980).
15. H.E. Newsom, W.M. White, K.P. Jochum and A.W. Hofmann. Siderophile and chalcophile element abundences in oceanic basalts, Pb isotope evolution and growth of the Earth's core, Earth Planet. Sci. Letters 80, 299-313 (1986).
16. S.V. Rasskazov. The Udokan basaltoids, Baikal Rift Zone. Novosibirsk, Nauka Publishers (1985). (in Russian)
17. S.V. Rasskazov. Magmatism related to the Eastern Siberia rift system and the geodynamics, Bull. Centres Rech. Explor.-Prod. Elf. Aquitaine 18, 437-452 (1994).
18. S.V. Rasskazov, A. Boven, L. Andre, J-P. Liegeois, A.V. Ivanov, L. Punzalan. Magmatic evolution of the northeastern Baikal Rift System, Petrology (translated from Russian) 5, N2 (1997).
19. S.V. Rasskazov, S.A. Bowring, N. Harris, J.F. Luhr, A.V. Ivanov. Variations of the Pb, Sr and Nd isotopic compositions in differentiated alkaline series from the Udokan volcanic field of the Baikal rift system (abs.) XIV symposium on isotope geochemistry. 198-199 (1995). (in Russian).
20. S.V. Rasskazov, A.V. Ivanov, I.S. Brandt, S.B. Brandt. Migrating Late Cenozoic volcanism of the Udokan field in the structures of the Baikal and Olekma-Stanovaya systems, Dokladi Academii Nauk. (1997) (in Russian) (in press).
21. S.V. Rasskazov, N.A. Logatchev, A.V. Ivanov. Correlations of Late Cenozoic tectonic and magmatic events in the Baikal rift system and those in the southeastern Eurasian plate, Geotectonica. (1997) (in Russian) (in press).
22. S.I. Sherman and K.G. Levi. Transform faults of the Baikal rift zone and seismicity of there terminations, Tectonics and seismicity of the continental rift zones. Nauka publishers, Moscow, pp. 7-18 (1978). (in Russian).
23. S.-S. Sun and W.F. McDonough. Chemical and isotopic systematics of oceanic basalts: implications for mantle composition and processes. In: Magmatism in the ocean basins. Geological Society Special Publication 42. A.D. Sounders, M.J. Norry (Eds). pp. 313-345 (1989).
24. S.R. Taylor and S.M. McLennan. The continental crust: its composition and evolution. Blackwell Scientific Publications (1985).
25. S.H. Yoon and S.K. Cough. Regional strike slip in the eastern continental margin of Korea and its tectonic implications for the evolution of Ulleung Basin, East Sea (Sea of Japan), Geol. Cos. Am. Bull. 107, 83-97 (1995).
26. R.E. Zartman, K. Futa, Z.C. Peng. A comparison of Sr-Nd-Pb isotopes in young and old continental lithospheric mantle: Patagonia and eastern China, Australian J. Earth Sci. 38, 545-557 (1991).
27. M. Zhang, P. Suddaby, R.N. Thompson, M.F. Thirlwall and M.A. Menzies. Potassic volcanic rocks in NE China: Geochemical constraints on mantle source and magma genesis, Jour. Petrol 36, 1275-1303 (1995).
28. Y.A. Zorin, V.M. Kozhevnikov, M.R. Novoselova and E.K. Turutanov. Thickness of the lithosphere beneath the Baikal rift zone and adjacent regions, Tectonophysics 168, 327-337 (1989).

Proc. 30th Int'l. Geol. Congr., Part 15, pp. 169 – 183
Li et al. (Eds.)
©VSP 1997

Geochemistry of the 2.4-2.3 Ga Kuandian Complex, Sino-Korean Craton: A Paleoproterozoic continental rift?

MIN SUN[1], LIFEI ZHANG[1,2], JIAHONG WU[3] AND ROB KERRICH[4]

[1]*Department of Earth Sciences, University of Hong Kong, Hong Kong;*
[2]*Department of Geology, Peking University, Beijing, China;*
[3]*Shenyang Institute of Geology and Mineral Resources, Shenyang, China;*
[4]*Department of Geological Sciences, University of Saskatchewan, Saskatoon, Saskatchewan, Canada S7N 0W0*

Abstract

The well preserved 2.3-2.4 Ga Kuandian Complex is located in a Proterozoic mobile belt bounded by Archean blocks of the northeastern Sino-Korea Craton. The Complex is mainly composed of amphibolites, gneisses, and layered granites. Geochemical studies show that the protoliths of the complex are associations of a bimodal magmatic suite with ε_{Nd} values between 0.26 and 1.94. The Kuandian amphibolites are depleted in Nb, Ta, and Ti, and enriched in K, Rb, and Pb, with pronounced depletion of Sr relative to Nd and Pb, Nb/La ratios are lower than 1 (0.3-0.8). The trace element patterns of the amphibolites are similar to those of continental flood basalts formed by the rifting of Gondwana, such as the Karoo and Tasmania basalts, featuring crustal contamination. Trace element modeling indicates that Kuandian gneisses and granite could be produced by extensive fractional crystallization from a magma with a composition similar to those of the amphibolites. This study indicates that the evolution of the Sino-Korean Craton is in contrast with that of many other Precambrian continents, e.g. North America, Europe, and Australia, in terms of vigorous magmatic activities between 2.3 and 2.4 Ga in the Sino-Korea Craton.

keywords: Kuandian Complex, Paleoproterozoic, geochemistry, continental flood basalt, northeastern China

INTRODUCTION

The Kuandian Complex, located in the Liaodong (east Liaoning) Peninsula, northeastern China, is a well preserved Paleoproterozoic complex of metamorphosed extrusive and intrusive rocks, formed by 2.3-2.4 Ga magmatic activity [38]. There are contrasting interpretations as to the petrogenesis and tectonic setting of this complex. The complex was traditionally considered as the basal part of the thick Proterozoic "Liaohe" Group [25]. Zhang [47,48] used "Liaojitite" to distinguish these rocks from the majority of sediments of the Liaohe Group and considered that these meta-igneous rocks formed from genetically related magmas in a eugeosyncline environment, whereas those sediments formed in a miogeosyncline. Jiang [24] used "Kuandian Group" for the Complex and suggested the protoliths of the complex to be oceanic volcanic rocks associated with some sediments. Alternatively, Bai et al. [5] proposed that the complex is a product of subduction zone magmatism of a convergent margin setting. This study is focused on the geochemistry of the Kuandian Complex in order to evaluate its petrogenesis and tectonic setting. From this basis this paper provides additional data to place constraints on its significance and context in the larger scale evolution of the Sino-Korean Craton in early Precambrian times.

The majority of the continental lithosphere formed early in the history of the Earth [e.g. 2,3], possibly with two important crustal accretion periods in the Precambrian at 2.6-2.8 and 1.7-1.9 Ga, respectively [7]. The apparent lack of major magmatic activity between 2.4 and 2.0 Ga has led to suggestions that there was a magmatic quiescence during the time interval in the Paleoproterozoic [8]. Some ages falling in the time interval have been reported since 1990 for rocks in west Africa and China [1, 6, 34, 38]. Accordingly, further studies of rocks with ages between 2.4 and 2.0 Ga, such as the Kuandian Complex, may shed light in understanding the history of crustal growth and earth evolution.

GEOLOGICAL BACKGROUND

The Kuandian Complex occurs in an EW-NEE trending Paleoproterozoic mobile belt bounded by Archean blocks in the Sino-Korean Craton (Fig. 1). This belt extends to the Korean peninsular in the east and is terminated to the west by the Tan-Lu (Tancheng-Lujiang) Fault. Early researchers assigned the medium to high grade gneisses and amphibolites of the belt to the lower Liaohe Group [e.g. 25]. Zhang [47,48] used the term Liaojitite Suite for a sequence of metavolcanics, granites and overlying turbidites,

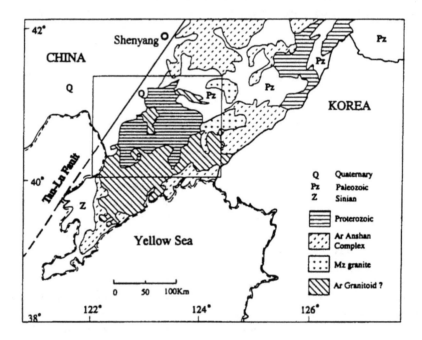

Figure 1. Simplified geological map of eastern Liaoning Province and adjacent areas, China. Jilin Province is north of 41°N and east of 126°E. Figure 2 is outlined by a rectangle.

which occur at the bottom of the lower Liaohe Group. Jiang [24] proposed Kuandian Group for the medium to high grade metamorphic rocks, distinguished from the Liaohe Group, based on different and discontinuous lithologies of the Liaohe Group in the north and south regions of Liaoning-Jilin provinces. The Kuandian Group unconformably overlies the Archean Anshan Group, and in turn is unconformably overlain by the Liaohe Group [24]. In this paper we use the term Kuandian Complex for the amphibolites, gneisses and associated interlayered granites.

The Kuandian Complex is composed principally of gneisses, amphibolites and granites, all collectively metamorphosed to the amphibolite facies. Figure 2 shows sample localities. The amphibolites are usually intercalated with, or included as lenticular inclusions in, the gneisses. Amphibolite layers are metres to tens of metres thick, concordant with foliation in the surrounding gneisses. Mineralogically the amphibolites consist of medium to fine grained blasts of hornblende + plagioclase +/- diopside + quartz. Dimensional alignment of prismatic hornblende defines lineation and foliation. No relict igneous textures are observed. According to M. Sun et al. [38], the protoliths of the amphibolites are considered as tholeiitic basalts.

The Kuandian gneisses include medium and fine grained varieties (the latter are usually termed leptite and leptynite in China), usually intercalated with amphibolites. They are composed of either medium or fine grained blasts of biotite + hornblende + albite + quartz + potassium feldspar. Relict porphyritic texture may be observed in biotite-albite-

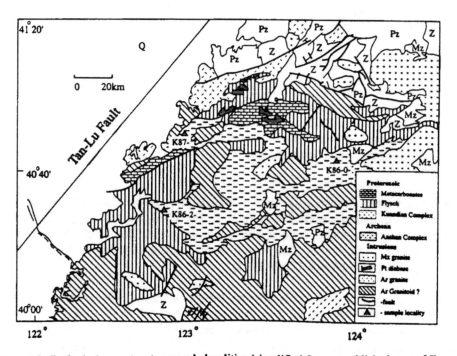

Figure 2. Geological map showing sample localities (simplified from unpublished map of Jiang, 1987).

fine-grained gneisses. The protolith of the gneisses are probably dacitic rhyolite or rhyolitic dacite [47,48].

Granites in the Kuandian Complex are usually layered, and contain inclusions of the amphibolites. The relationships between the granites and gneisses/amphibolites are obscured by high intensities of deformation and metamorphism. Zhang [47,48] suggested conformable contacts between these rock types. The main constituent minerals of the granites are microcline, oligoclase, hornblende and quartz. Accessory minerals include magnetite and euhedral zircons with length/width ratios of 2-3. The Kuandian granites show medium-fine grained equigranular blastic textures; occasionally porphyroblastic textures are observed [38]. The Kuandian granites were once termed lineated hornblende migmatite [23,24], and now tend to be considered as granite with a magmatic origin [38,48].

Previous isotope geochronological data for rocks from the Kuandian Complex scatter, due to a combination of high grade metamorphism and large analytical errors. K-Ar ages for the gneisses cluster at 1.9 to 1.5 Ga. Rb-Sr isochron ages for fine-grained gneisses vary between 2.2 and 1.9 Ga [46]. Upper intercept U-Pb ages of zircons from the granites are 2.3 to 1.8 Ga [48]. Recent precise isotopic data reveal a formation age of 2.4 to 2.3 Ga for the Kuandian Complex, based on a 2.4+/-0.1 Ga Rb-Sr isochron for medium grained gneisses [36]; a 2.3+/-0.1 Ga Sm-Nd isochron for the Kuandian amphibolites and granites; and a 2.25+/-0.05 Ga upper intercept U-Pb zircon age for the Kuandian granites [38]. Nd depleted mantle model ages are 2.46-2.75 Ga for the Kuandian amphibolites, and 2.36-2.53 Ga for the Kuandian granites [38].

RESULTS

Major element analyses were carried out with an XRF spectrometer. Trace element analyses were performed on a Perkin Elmer Elan 5000 ICP-MS at the University of Saskatchewan. To ensure complete digestion of zircon and other accessory minerals, a Na_2O_2 sinter method was used for high field strength elements, and $HF-HNO_3$ digestion was used for other trace elements. Analysis followed a procedure described by Sun and Kerrich [39].

Major and trace element data for the amphibolites, gneisses and granites from the Kuandian Complex are presented in Table 1. On a SiO_2 vs. K_2O+Na_2O diagram, the samples show subalkaline features and a bimodal distribution is pronounced for the three major rock types of the Kuandian Complex, lacking intermediate compositions in the range SiO_2 = 58-66 wt% (Table 1)[35].

The SiO_2 contents of the amphibolites vary between 47% and 54%, excepting for one of 58%. In an AFM diagram, the Kuandian amphibolites fall in tholeiite field [38]. The SiO_2 contents of the Kuandian gneisses vary between 66% to 88%, falling in dacite and rhyolite fields in the total alkalis versus silica classification diagram for volcanic rocks proposed by Cox et al. [9]. The SiO_2 contents of the Kuandian granites are concentrated between 73% to 75%, while contents of alkali oxides are 8-9%. On an Ab-An-Or

Table 1. Major and trace element data for the Kuandian Complex

	K86031	K86083	K86244	K86245	K86246	K86247	K87087	K87137
SiO2	47.4	50.4	48.9	53.2	49.1	54.3	52.2	58.2
TiO2	1.27	1.29	1.05	1.07	0.95	0.65	0.79	0.44
Al2O3	13.0	14.7	15.3	14.5	14.4	15.6	14.1	13.3
Fe2O3	12.4	14.1	12.7	12.3	11.8	11.3	12.9	8.44
MnO	0.17	0.25	0.18	0.23	0.20	0.11	0.21	0.18
MgO	6.69	6.04	6.65	6.17	8.99	3.10	6.25	6.71
CaO	8.96	9.65	10.3	8.38	11.3	6.38	10.1	7.49
Na2O	2.00	2.41	2.38	2.43	2.21	2.16	0.90	2.79
K2O	1.06	1.07	2.46	1.56	0.96	4.73	0.34	1.15
P2O5	0.15	0.11	0.10	0.14	0.08	0.14	0.11	0.28
H2O	nd	0.85	0.42	nd	0.85	nd	nd	nd
V	300.	282.	185.	266.	176.	310.	307.	129.
Rb	48.7	91.5	132.	31.1	29.5	28.9	17.0	64.3
Sr	383.	293.	226.	204.	212.	261.	244.	367.
Y	20.7	15.4	18.0	20.4	16.1	23.4	16.8	14.0
Zr	66.8	56.0	75.0	76.2	60.1	90.7	74.1	109.
Nb	6.16	3.12	5.02	5.83	5.01	7.14	4.32	5.10
Ba	1070.	482.	210.	215.	137.	156.	101.	426.
Hf	1.91	1.62	2.04	2.47	1.62	2.87	1.83	3.01
Ta	0.37	0.20	0.27	0.35	0.22	0.47	0.27	0.40
Pb	6.95	13.5	2.47	7.95	3.63	6.15	13.3	11.1
Th	1.59	1.55	2.06	2.98	1.62	1.30	1.70	9.49
U	0.45	0.30	0.38	0.70	0.32	0.40	0.41	1.27
La	9.80	6.03	8.24	10.7	5.97	15.5	10.7	17.8
Ce	21.8	14.0	19.1	25.5	15.0	35.5	23.4	41.6
Pr	2.95	1.96	2.74	3.23	2.22	4.54	2.99	4.12
Nd	12.7	8.86	12.4	13.7	10.5	19.3	12.4	15.4
Sm	3.39	2.30	3.08	3.66	2.75	5.01	2.77	3.32
Eu	1.28	0.82	0.99	1.18	0.90	1.76	0.94	0.96
Gd	4.32	2.81	3.48	4.47	3.17	5.81	3.19	3.14
Tb	0.66	0.46	0.56	0.67	0.53	0.89	0.48	0.43
Dy	4.34	3.11	3.82	4.23	3.42	5.62	3.32	2.72
Ho	0.92	0.64	0.72	0.86	0.67	1.12	0.67	0.54
Er	2.70	1.86	2.09	2.37	1.89	3.08	2.05	1.51
Tm	0.39	0.28	0.31	0.34	0.27	0.45	0.29	0.21
Yb	2.49	1.62	1.88	2.14	1.65	2.81	1.96	1.46
Lu	0.38	0.25	0.26	0.31	0.24	0.41	0.28	0.21

	K86110	K86112	K86195	K86197	K86199	K86202	K870191	K86027
SiO2	66.5	67.6	80.6	74.8	75.6	72.0	74.4	74.7
TiO2	0.47	0.44	0.24	0.18	0.29	0.28	0.38	0.27
Al2O3	16.9	15.6	8.89	13.2	12.2	13.7	11.9	11.4
Fe2O3	4.11	4.47	2.57	1.97	2.73	3.43	4.06	4.40
MnO	0.06	0.08	0.06	0.05	0.06	0.10	0.02	0.07
MgO	1.81	1.69	0.81	0.62	1.14	1.39	0.20	0.10
CaO	2.04	2.88	0.69	1.20	0.35	2.29	0.14	0.53
Na2O	1.76	1.92	1.53	3.56	1.61	3.20	1.46	3.30
K2O	4.20	4.10	3.91	3.75	4.01	2.51	6.81	5.16
P2O5	0.17	0.16	0.06	0.07	0.10	0.08	0.09	0.03
H2O	nd	nd	nd	nd	nd	nd	nd	0.35
V	52.59	54.7	32.9	19.4	51.1	27.0	20.6	27.0
Rb	174.	199.	155.	104.	284.	177.	202.	149.
Sr	208.	287.	113.	168.	108.	220.	49.1	63.6
Y	38.2	30.4	14.8	23.7	11.2	15.1	21.1	48.6
Zr	218.	196.	99.2	148.	101.	169.	244.	367.
Nb	20.7	16.9	8.32	11.6	10.5	13.6	13.1	20.6
Ba	722.	883.	1013.	962.	1325.	833.	728.	803.
Hf	6.33	5.11	2.62	3.93	2.49	4.49	6.64	8.85
Ta	1.40	1.13	0.58	1.25	0.74	1.02	0.85	1.23

M. Sun et al.

Table 1 (continued)

	K86110	K86112	K86195	K86197	K86199	K86202	K870191	K86027
Pb	7.79	12.1	25.6	20.2	17.2	25.5	5.16	5.76
Th	24.3	19.9	8.64	26.5	13.3	16.4	13.4	18.4
U	6.78	4.78	2.50	4.62	2.24	1.77	2.67	1.88
La	68.6	54.6	17.9	35.8	11.5	30.6	48.0	77.7
Ce	135.	105.9	38.8	77.2	48.7	89.8	100.	154.
Pr	14.5	12.8	3.75	7.76	2.80	6.23	11.5	14.6
Nd	49.3	45.2	13.8	25.3	9.97	21.6	39.1	51.4
Sm	9.41	8.30	2.41	4.58	2.01	3.82	5.93	8.34
Eu	1.09	1.05	0.50	0.63	0.36	0.60	1.11	1.04
Gd	8.09	7.24	2.29	3.84	2.04	3.26	4.58	6.79
Tb	1.12	0.94	0.33	0.54	0.27	0.41	0.58	0.96
Dy	6.98	6.08	2.31	3.25	1.81	2.70	3.03	6.70
Ho	1.40	1.16	0.48	0.65	0.40	0.52	0.51	1.27
Er	4.08	3.44	1.40	2.02	1.31	1.68	1.45	3.55
Tm	0.59	0.52	0.22	0.30	0.21	0.29	0.22	0.59
Yb	3.60	3.06	1.29	2.03	1.33	1.98	1.42	3.67
Lu	0.49	0.42	0.21	0.30	0.21	0.31	0.21	0.63

	K86086	K86088	K86090	K86176	K86178	K86190	K86191	K86193
SiO2	72.8	73.9	74.1	73.4	74.2	74.3	73.8	75.2
TiO2	0.31	0.30	0.30	0.28	0.30	0.27	0.58	0.29
Al2O3	11.8	11.7	11.6	12.2	12.1	12.3	12.4	12.2
Fe2O3	4.80	4.50	4.70	4.59	4.34	3.44	3.64	3.61
MnO	0.09	0.07	0.07	0.06	0.03	0.05	0.04	0.03
MgO	0.04	0.06	0.05	0.10	0.18	0.11	0.06	0.09
CaO	1.28	0.63	1.06	0.24	0.30	0.90	1.03	0.19
Na2O	3.17	3.37	3.95	3.43	3.60	3.45	3.63	3.45
K2O	5.68	5.34	4.18	5.21	4.83	4.77	4.90	4.87
P2O5	0.04	0.03	0.03	0.03	0.03	0.03	0.03	0.03
H2O	0.21	0.37	nd	nd	nd	nd	nd	nd
V	34.1	29.7	22.3	3.16	1.89	3.03	1.54	8.69
Rb	219.	201.	159.	147.	164.	194.	183.	158.
Sr	83.9	75.9	107.	45.0	26.3	99.6	79.1	51.1
Y	57.5	48.1	52.6	35.8	46.6	52.4	59.1	49.4
Zr	263.	263.	365.	313.	307.	275.	324.	329.
Nb	25.9	24.2	25.2	21.4	21.6	22.5	24.3	22.2
Ba	1093.	936.	681.	985.	685.	879.	968.	863.
Hf	7.73	8.68	9.35	8.51	7.85	7.55	8.68	8.49
Ta	1.84	1.68	1.75	1.42	1.39	1.85	1.71	1.58
Pb	9.09	9.85	6.56	5.36	5.36	8.26	6.24	12.4
Th	14.8	19.4	22.0	19.7	16.6	26.5	17.6	20.2
U	1.56	1.95	0.96	1.79	2.29	3.23	3.27	2.57
La	40.4	53.0	97.1	49.2	52.2	85.0	66.1	52.8
Ce	86.4	106.	208.	81.6	98.3	156.	121.	96.1
Pr	10.6	12.2	21.9	13.6	14.0	18.8	13.15	12.9
Nd	43.9	47.4	78.1	50.1	53.0	66.0	47.9	45.4
Sm	9.43	9.15	12.3	9.41	10.6	10.8	10.1	8.39
Eu	1.49	1.47	1.47	1.62	1.84	1.57	1.58	1.26
Gd	9.94	9.03	10.5	7.57	9.36	10.3	10.3	6.58
Tb	1.62	1.32	1.50	1.14	1.30	1.44	1.56	0.96
Dy	10.9	8.79	9.65	7.13	8.14	8.87	10.4	5.72
Ho	2.26	1.77	1.83	1.37	1.62	1.76	2.07	1.15
Er	6.82	5.39	5.46	4.05	4.50	5.17	6.04	3.27
Tm	0.98	0.80	0.76	0.60	0.74	0.75	0.85	0.49
Yb	6.45	5.48	4.71	4.11	4.55	4.97	5.44	3.18
Lu	0.91	0.73	0.67	0.65	0.75	0.73	0.80	0.47

Amphibolite samples: K86031, K86083, K86244, K86245, K86246, K86247, K87087, K87137.

Gneiss samples: K86110, K86112, K86195, K86197, K86199, K86202, K870191.

Granite samples: K86027, K86086, K86088, K86090, K86176, K86178, K86190, K86191, K86193.

diagram, the Kuandian granites plot in the granite field [38].

Correlation of major elemental compositions for the three rock types is very obvious, as shown by a linear trend in a $(MgO+FeO)$- (K_2O+Na_2O) - CaO diagram (Fig. 3), implying a possible genetic relationship.

In a primitive mantle normalized diagram, the Kuandian amphibolites show pronounced depletions in Nb, Ta, and Ti, and enrichments of K, Rb, Ba, and Pb (Fig. 4). Chondrite normalized REE patterns of the amphibolites are slightly LREE enriched, $(La/Yb)_N$ = 2.7-3.8. Samples of the amphibolites either have no Eu anomaly or show a slight negative Eu anomaly.

The gneisses are enriched in K, Rb, Th, U, Cs, and Pb, and depleted in Nb, Ta, Ti, and Sr (Fig. 4). Chondrite normalized REE patterns for the gneisses are highly LREE enriched, $(La/Yb)_N$=2.7-21.7, with pronounced negative Eu anomalies, Eu/Eu^*=0.38-0.84.

The Kuandian granites are strongly depleted in Nb, Ta, Sr, and Ti, and enriched in Rb, Th, K, and Pb (Fig. 4). The trace elemental patterns of the granites are similar to those of A-type granites in the Topsails terrane in Newfoundland, Canada [44]. The Kuandian granites are highly LREE enriched, $(La/Yb)_N$ = 2.7-17.2, with striking negative Eu anomalies, Eu/Eu^* = 0.36-0.59.

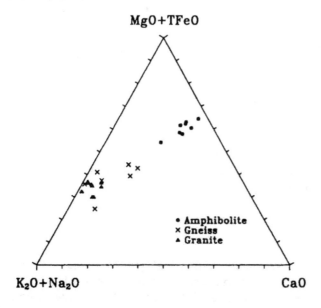

Figure 3. MgO+ΣFeO - K₂O+Na₂O - CaO diagram for the amphibolites, gneisses, and granites from the Kuandian Complex.

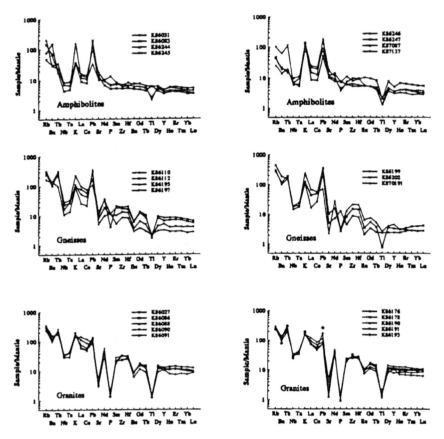

Figure 4. Primitive mantle normalized trace element diagrams for the amphibolites, gneisses, and granites from the Kuandian Complex. Normalized values are from Sun and McDonough (1989).

In summary, although the three major rock types of the Kuandian Complex have different trace element contents, they have similar patterns, i.e. depleted in Nb, Ta, and Ti, and enriched in K, Rb, and Pb. These features could also imply a genetic relation for the Kuandian Complex.

DISCUSSION

1. amphibolites
Based on lithological association, Zhang et al. [48] suggested that the protoliths of the Kuandian amphibolites are mantle-derived tholeiitic basalts formed in an extensional environment. However, Jiang et al. [24] considered an oceanic volcanic-sedimentary sequence for the amphibolites, largely based on their field occurrence. Bai et al. [5] proposed that the protoliths of the amphibolites are arc volcanic rocks.

The Kuandian amphibolites are enriched in LILE, but are depleted in Nb, Ta, and Ti. Nb/La ratios for the Kuandian amphibolites are all smaller than 1 (0.3-0.8). Depletion of

Nb and Ta and enrichment of K and Pb are common features for arc basalts and some flood basalts contaminated by continental materials [4, 40, 42]. However, arc basalts are very different in Sr contents from the continental flood basalts: the former is enriched and the latter is depleted in Sr [40]. The enrichment of Sr relative to Pb and Nd has been considered as a significant character for the arc basalts [17, 29]. The Kuandian amphibolites are depleted in Sr relative to Pb and Nd. Lacking or slight negative Eu anomalies are also inconsistent with arc basalts. Trace element patterns of the Kuandian amphibolites are similar to those of the Karoo, Tasmania, and Parana basalts, which are genetically related to the Gondwana continental breakup and show crustal contamination (Fig. 5). Higher K in the Kuandian amphibolites could be due to metamorphism/metasomatism, and lower Ti could be possibly due to crystallization of ilmenite. Therefore we infer that the protoliths of the Kuandian amphibolites are contaminated continental flood basalts.

Mg' values [$Mg/(Mg+Fe^{2+})$] for the Kuandian amphibolites vary between 0.50 and 0.65, except one at 0.39. The Kuandian amphibolites also have low Ni (42-109 ppm) and low Cr contents (97-534 ppm)[38]. These features are common for continental flood basalts [45 and reference therein], indicating that the protoliths of the Kuandian amphibolites were derived from evolved magmas.

In a ε_{Nd} - t diagram (Fig. 6), the Kuandian Complex is less depleted than Archean mantle-derived rocks in the Sino-Korean Craton. Initial $^{143}Nd/^{144}Nd$ and $^{87}Sr/^{86}Sr$ ratios for the Kuandian amphibolites are similar to those of the Karoo, Parana, Ferrar, and Tasmanian basalts. These basalts from the southern hemisphere have higher radiogenic Pb than the basalts from the Pacific and from the north Atlantic Ocean, and hence show

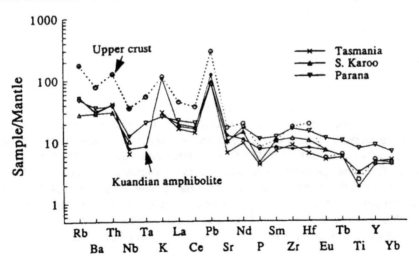

Figure 5. Primitive mantle normalized trace element diagrams for the Kuandian amphibolites (average value of this study), S. Karoo (data from Duncan et al., 1984; Ellam and Cox, 1991), Parana (data from Hawkesworth et al., 1988; 1992), and Tasmania basalts (data from Hergt et al., 1989). The upper crust is also plotted for comparison, data from Taylor and McLennan (1985).

M. Sun et al.

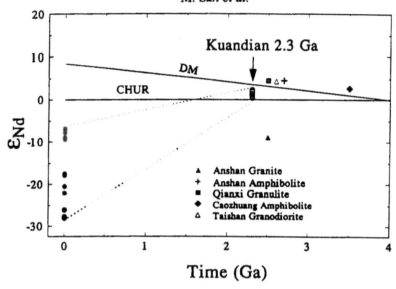

Figure 6. ε_{Nd} - time diagram for the Kuandian amphibolites and granites (Data from M. Sun et al., 1993). Data for Archean rocks of the Sino-Korean Craton are plotted for comparison (Jahn and Zhang , 1984; Jahn et al., 1984; Jahn et al., 1987; Jahn et al., 1988; Jahn, 1990; Jahn and Ernst, 1990; Qiao et al., 1990).

Dupal anomalies [11,12,14]. The Dupal anomaly is usually interpreted as a result from the contamination of a depleted mantle by the subduction of sedimentary materials under Pangea [40,49]. Menzies and Kyle [27] proposed that the Ferrar and other southern hemisphere flood basalts with the Dupal anomaly were derived from a depleted mantle (DMM) mixed with recycled crustal components. In $^{207}Pb/^{204}Pb$ - $^{206}Pb/^{204}Pb$ and $^{208}Pb/^{204}Pb$ - $^{206}Pb/^{204}Pb$ diagrams, a plagioclase separate from the Kuandian amphibolites also shows a signature of high radiogenic Pb [35,38]. We consider this as evidence for crustal contamination.

Hart [14] discussed the Dupal anomaly and pointed out that the anomaly mainly occurs near 30°S in the southern hemisphere, possibly since the Archean. Some recent studies also show a Dupal signature in the northern Hemisphere, e.g. the Philippines, Japan Sea, and South China Basin [28,33,43]. If a Dupal anomaly exists for the 2.3-2.4 Ga Kuandian Complex, problems will arise for the geographical location of the Sino-Korean Craton in the Paleoproterozoic and for the tectonic significance of the anomaly. Notwithstanding all these interesting problems, we infer that magma for the Kuandian amphibolites originated from a contaminated mantle possibly due to crustal recycling. This may signify that a modern style mantle-crust recycling has existed at least for 2.3-2.4 Ga.

2. amphibolites, granites and gneisses
Major element data and trace element patterns suggest that the Kuandian amphibolites, gneisses, and granites are genetically related. Little variation exists for the initial ε_{Nd} values of the Kuandian amphibolites (0.70-1.94) and Kuandian granites (0.26-1.85) and

data fall on a Sm-Nd isochron [38]. This also supports the above interpretation. The Kuandian gneisses and granites fall in within-plate fields in Rb - Y+Nb and Nb - Y diagrams [38]. Therefore we consider that the bimodal magmatic activity developed in an aulacogen environment.

The protoliths of the gneisses and granites resulted either from fractional crystallization of a basaltic magma, or from partial melting of solidified basalts underplated at the base of the crust. Partial melting cannot explain the depletion of Sr and Eu in the gneisses and granites, because plagioclase has very high contents of Sr and Eu, and it should not be a residual phase. Nb/Yb ratios positively covary with ratios of La/Yb for the Kuandian amphibolites, gneisses, and granites, which can be attributed to fractional crystallization (Fig. 7).

The chondrite normalized ratios of $(Nb/La)_N$ and initial ε_{Nd} for the Kuandian amphibolites, gneisses, and granites are very consistent (0.3-0.9, and 0.26-1.94, respectively), indicating that the members of the Kuandian Complex underwent a similar extent of contamination, because the Nb/La ratio and ε_{Nd} are sensitive to contamination. Same extent of contamination is difficult to achieve during magma ascending or different eruptions. This suggests that the contamination of the Kuandian Complex happened in the magma source, i.e. crustal recycling in the mantle or in the magma chamber at the base of the crust. The values of ε_{Nd} for the Kuandian Complex are all greater than 0, much higher than those of the most likely local crustal contaminant, e.g. the Archean Anshan granites with ε_{Nd} value of -10. This also eliminates a significant contamination by local crustal materials.

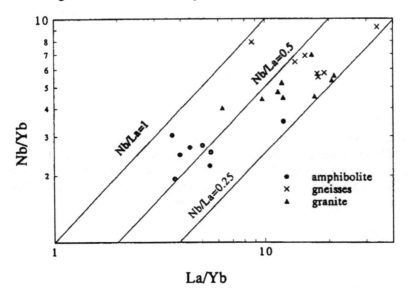

Figure 7. Nb/Yb vs. La/Yb diagram for the amphibolites, gneisses, and granites from the Kuandian Complex.

Table 2. Fractional crystallization modeling for the Kuandian Complex

Elements (ppm)	amphibolite (avg)	gneiss (avg, measured)	gneiss (calculated)	granite(avg, measured)	granite (calculated)
Rb	34.2	185.	185.	177.	178.
Ba	207.	924.	924.	893.	893.
Th	3.42	17.5	17.2	19.3	20.0
Nb	8.46	13.5	13.6	22.9	21.8
K	26502	35550	112062	42278	100005
La	12.1	38.1	37.9	61.7	60.8
Sr	258.	165	161.	70.1	69.8
Sm	3.50	5.21	8.22	9.81	14.5
Zr	82.1	168.	256.	318.	316.
Eu	1.15	0.76	0.79	1.48	1.44
Yb	2.01	2.10	2.21	4.76	4.46
f = redidual magma (%)			15		15
crystallized phases %			Ol(22) Cpx (30) Pl (30) Gt(18)		Ol(20) Cpx (20) Pl(50) Gt(10)

Because of the above reasons, there is no need to use the AFC (assimilation with fractional crystallization) model for the petrogenesis of the Kuandian Complex. Accordingly we used average composition of the Kuandian amphibolites as the composition of parent magma, and calculated fractional crystallization products by choosing different minerals and different extent of crystallization (Table 2). Intensive fractional crystallization of olivine, clinopyroxene, plagioclase and other minor phases, possibly at the lower crustal level, can produce rocks with compositions similar to those of the Kuandian gneisses and granites.

The extensive magmatic activity in the Sino-Korean Craton is in sharp contrast with a magmatically quiet period between 2.4 and 2.0 Ga in North America, Europe, and Australia [7, 26, 30, 31]. Association of bimodal volcanic-arkose-conglomarate (BVAC) is common for Proterozoic aulacogens, e.g. the 1.8-1.9 Ga Dewars and Waterberg in South Africa, and Pahrump and Uinta Mountain Groups in North America [7]. The Kuandian Complex is overlain by turbidite, and the association was named Liaojitite by Zhang [47,48]. The Kuandian Complex and the overlying turbidite could be a similar counterpart of a 2.3-2.4 Ga BVAC. Magmatic activity in this time period has been also reported in the Zhongtiao Mountain area, Shanxi Province, China [34].

CONCLUSIONS

The magma for the Kuandian Complex was derived from a mantle source contaminated by crustal material, possibly due to a crustal recycling before 2.4 Ga. The protoliths of the amphibolites are continental flood basalts, comparable to flood basalts formed during the break-up of Gondwana. The gneisses and granites are the results of an extensive fractional crystallization of the parental magma with a composition of the amphibolites. The Kuandian Complex formed in an aulacogen tectonic environment 2.3

to 2.4 Ga before present, in sharp contrast with a magmatically quiet period for many other cratons.

Acknowledgments

The first author is in debt to Professor Jiang Chun Chao for his field guidance during sample collection. J. Fan is thanked for analytical assistance. The research was supported by a CRCG grant from the University of Hong Kong and the George McLeod endowment to the Department of Geological Sciences, University of Saskatchewan. Croucher Foundation scholarship to L.F. Zhang is gratefully acknowledged.

REFERENCES

1. W. Abouchami and M. Boher. A major 2.1 Ga event of mafic magmatism in West Africa: An early stage of crustal accretion, *Jour. Geophys. Res.* 95, 17605-17629 (1990).
2. R.L. Armstrong. A model for the evolution of strontium and lead isotopes in a dynamic earth, *Rev. Geophys.* 6, 175-199 (1968).
3. R.L. Armstrong. Radiogenic isotopes: the case for crustal recycling on a near-steady-state no-continental growth earth, *Phil. Trans. R. Soc. London* 301, 443-472 (1981).
4. N.T. Arndt and U. Christensen. The role of lithospheric mantle in continental flood volcanism: thermal and geochemical constraints, *Jour. Geophys. Res.* 97, 10967-10981 (1992) .
5. J. Bai, X.G. Huang, F.Y. Dai and C.H. Wu. *Precambrian crustal evolution of China.* Geological Publishing House, Beijing, China (1993).
6. M. Boher, W. Abouchami, A. Michard, F. Albarede and N.T. Arndt. Crustal growth in West Africa at 2.1 Ga, *Jour. Geophy. Res.* 97, 345-369 (1992).
7. K.C. Condie. *Plate Tectonic and Crustal Evolution*, Pergamon, (1989).
8. K.C. Condie. *Proterozoic Crustal Evolution*, Pergamon (1992).
9. K.G. Cox, J.D. Bell and R.L. Pankhurst. *The interpretation of igneous rocks.* George, Allen and Unwin, London (1979).
10. A.R. Duncan, A. J. Erlank and J.S. March. Regional geochemistry of the Karoo igneous province. *Special Pub. Geol. Soc. South Africa* 13, 355-388 (1984).
11. B. Dupre and C.J. Allegre. Pb-Sr-Nd isotopic correlation and the chemistry of the North Atlantic mantle, *Nature* 286, 17-21 (1980).
12. B. Dupre and C.J. Allegre. Pb-Sr isotope variation in Indian Ocean basalts and mixing phenomena, *Nature* 303, 142-146 (1983).
13. R.M. Ellam and K.G. Cox. An interpretation of Karoo picrite basalts in terms of interaction between asthenospheric magmas and the mantle lithosphere. *Earth Planet. Sci. Lett.* 105, 330-342 (1991).
14. S.R. Hart. A large-scale isotope anomaly in the Southern Hemisphere mantle, *Nature* 309, 753-757 (1984).
15. C.J. Hawkesworth, M. Mantovani and D. Peate. Lithosphere remobilization during Parana CFB magmatism. In: *Oceanic and Continental Lithosphere: Similarities and differences.* M. A. Menzies and K. G. Cox (Eds.). pp. 205-224, Oxford University Press, Oxford (1988).
16. C.J. Hawkesworth, K. Gallagher, S. Kelley, M. Mantovani, D.W. Peate, M. Regelous and N.W. Rogers. Parana magmatism and the opening of the south Atlantic, In: *Magmatism and*

M. Sun et al.

The Causes of Continental Break-up. B. C. Storey, T. Alabaster and R. J. Pankhurst (Eds.), pp.221-240. Geol. Soc. Spec. Pub. 68 (1992).

17. J.M. Hergt, B.M. Chappell, M.T. McCulloch, I. McDougall and A.R. Chivas. Geochemical and isotopic constraints on the Origin of the Jurassic dolerites of Tasmania. *Jour. Petrol.* **30**, 841-883 (1989).

18. B-M. Jahn and Z-Q. Zhang. Archean granulite gneisses from eastern Hebei Province, China: rare earth geochemistry and tectonic implications, *Contrib. Mineral. Petrol.* **85**, 224-243 (1984).

19. B-M. Jahn, B. Auvray, J. Cornichet, Y-L Bai, Q-H. Shen and D-Y. Liu. 3.5 Ga old amphibolites from eastern Hebei province, China: field occurrence, petrography, Sm-Nd isochron age and REE geochemistry, *Precam. Res.* **34**, 311-346 (1987).

20. B-M. Jahn, B. Auvray, Q-H. Shen, D-Y. Liu, Z-Q. Zhang, Y-J. Dong, X-J. Ye, J. Cornichet and J. Mace. Archean crustal evolution in China: the Taishan complex, and evidence for juvenile crustal addition from long-term depleted mantle, *Precam. Res.* **38**, 381-403 (1988).

21. B-M. Jahn. Early Precambrian basic rocks of China. In: *Early Precambrian Basic Magmatism.* R. P. Hall and D. J. Hughs (Eds.) pp. 294-316. Blackie, Glasgow. (1990).

22. B-M. Jahn and W.G. Ernst. Late Archean Sm-Nd isochron age for mafic-ultramafic supracrustal amphibolites from the northeastern Sino-Korean Craton, China. *Precam. Res.* **46**, 295-306 (1990).

23. C-C. Jiang. On Precambrian stratigraphy and comparison of eastern Liaoning: Deliberation on usage of "Liaohe Group". *Bull. Chinese Acad. Geol. Sci.* **9**: 157-167 (1984).

24. C-C. Jiang. *Precambrian geology of eastern Liaoning and Jilin.* Science and Technology Press of Liaoning, Shenyang (1987).

25. Liaoning Geological Bureau. *Geological Memoir of Liaoning.* Geological Publishing House (1989).

26. M.T. McCulloch. Sm-Nd isotopic constraints on the evolution of Precambrian crust in the Australian continent. In: *Proterozoic Lithospheric Evolution,* A. Kroner (Ed.). AGU Geodynamics, Ser., 17, Washington, D. C. (1987).

27. M.A. Menzies and P.R. Kyle. Continental volcanism: a crustal-mantle probe, In: *Continental Mantle.* M. A. Menzies(Ed.), 157-177, Oxford Science Publication (1990).

28. S.B. Mukasa, M. Robert and G.B. James. Pb-isotopic compositions of volcanic rocks in the West and East Philippine island arcs: presence of the Dupal isotopic anomaly, *Earth Planet. Sci. Lett.* **84**, 153-164 (1987).

29. E. Nakamura, M.T. McCulloch and S-S. Sun. The influence of subduction processes on the geochemistry of Japanese alkaline basalts, *Nature* **316**, 55-58 (1985).

30. B.K. Nelson and D.J. Depaolo. Rapid production of continental crust 1.7 to 1.9 b.y. ago: Nd isotopic evidence from the basement of the North American Mid continent, *Geol. Soc. Amr. Bull.* **86**, 746-754 (1985).

31. P.J. Patchett and N.T. Arndt. Nd isotopes and tectonics of 1.9-1.7 crustal genesis, *Earth Planet. Sci. Lett.* **78**, 329-338 (1986).

32. G-S. Qiao, M-G. Zhai and Y-H. Yan. Geochronology of Anshan-Benxi Archean metamorphic rocks in Liaoning Province, NE China. Sci. Geol. Sin., **2**, 158-165 (1990).

33. S. Sakes. Nd isotopic heterogeneity of the oceanic upper mantle, *Earth Sci.* **41**, 23-34 (1987).

34. D-Z. Sun, H-M. Li, Y-X. Lin, H-F. Zhou, F-Q. Zhao and M. Tang. Precambrian geochronology, chronotectonic framework and model of chronocrustal structure of the Zhongtiao Mountains, *Acta Geol. Sinica* 3 (1991).

35. M. Sun. *Consolidation and mantle evolution of the Sinokorean Craton in Early Precambrian time* [unpublished Ph.D. thesis]. University of British Columbia, Vancouver (1991).

36. M. Sun, J-H. Wu and C-C Jiang. Geochronological study of the Kuandian Complex. In: *Progress of Precambrian research in the eastern Liaoning area*. C-C. Jiang, J-H. Wu, G-S. Feng, G-Q. Liu, X-M. Kong and M. Sun (Eds.). pp. 19-102. Institute of Geology and Mineral Resources, Shenyang (1991).

37. M. Sun, R.L. Armstrong and R.St.J. Lambert. Petrochemistry and Sr, Pb, and Nd isotopic geochemistry of Early Precambrian rocks, Wutaishan and Taihangshan areas, China, *Precam. Res.* **56**, 1-31 (1992).

38. M. Sun, R.L. Armstrong, R. St J. Lambert, C-C. Jiang and J-H. Wu. Petrochemistry and Sr, Pb and Nd isotopic geochemistry of Paleoproterozoic Kuandian Complex, the eastern Liaoning Province, China, *Precam. Research* **62**, 171-190 (1993).

39. M. Sun and R. Kerrich. Rare earth element and high field strength element characteristics of whole rocks and mineral separates of ultramafic nodules in Cenozoic volcanic vents of southeastern British Columbia, Canada, *Geochim. Cosmochim. Acta* **59**, 4863-4879 (1995).

40. S.S. Sun and W.F. McDonough. Chemical and isotopic systematics of oceanic basalts: implications for mantle composition and processes, In: *Magmatism in the Ocean Basins*. A.D. Saunders and M.J. Norry (Eds.). pp. 313-345. Geol. Soc. Spec. Pub. 42 (1989).

41. S.R. Taylor and S.M. McLennan. *The Continental Crust: Its Composition and Evolution*, Blackwell Scientific Publications (1985).

42. M.F. Thirlwall, T.E. Smith, A.M. Graham, N. Theodorou, P. Hollings, J.P. Davidson and R.J. Arculus. High field strength element anomalies in arc lavas: source or process?, *Jour Petrol.* **35**, 819-838 (1994).

43. K. Tu, M.F.J. Flower, R.W. Carlson, G. Xie, C-Y. Chen and M. Zhang. Magmatism in the South China Basin 1. Isotopic and trace-element evidence for an endogenous Dupal mantle component, *Chem. Geol.* **97**, 47-63 (1992).

44. J.B. Whalen, K.L. Currie and B.W. Chappell. A-type granites: geochemical characteristics, discrimination and petrogenesis, *Contrib. Mineral. Petrol.* **95**, 407-419 (1987).

45. M. Wilson. *Igneous Petrology, A Global Tectonic Approach*. Chapman & Hall (1989).

46. Z-P. Zhao. *Precambrian crustal evolution of the Sino-Korean Platform*. Science Publication House, Beijing, China (1993).

47. Q-S. Zhang. *Geology and metallogeny of the Early Precambrian in China*. Jilin People's Press, Changchun (1984).

48. Q.S. Zhang. *Early crust and ore deposits in eastern Liaoning Peninsula*. Geological Publishing House (1986).

49. A. Zindler and S. Hart. Chemical geodynamics, *Ann. Rev.Earth Planet. Sci.* **14**, 493-571 (1986).

Proc. 30th Int'l. Geol. Congr., Part 15, pp. 185 – 198
Li et al. (Eds)
© VSP 1997

Geochemical Study of Eclogitic Mineral Inclusions from Chinese Diamonds

WUYI WANG, SHIGEHO SUENO
Institute of Geoscience, University of Tsukuba, Tsukuba 305, Japan
HISAYOSHI YURIMOTO, EIICHI TAKAHASHI
Earth and Planetary Sciences, Tokyo Institute of Technology, Tokyo 152, Japan

Abstract

Major and trace element geochemistry of eclogitic mineral inclusions from Chinese diamonds are reported in this study, for the first time. Bulk major element compositions of mantle eclogite, estimated from diamond inclusions, are very close to that of MORB. All the analyzed samples exhibit evident positive Eu anomalies. Estimated bulk trace element compositions of mantle eclogite are generally parallel to that of MORB, but with deviations like enrichment in LILE and depletion in HFSE. It is proposed that the formation of mantle eclogite could be closely related to recycling of ancient oceanic crust. Other processes like (1) metasomatism by incompatible trace element rich melts; or (2) remelting and interaction with mantle peridotite, may also be involved. Coexisting of olivine with eclogitic mineral inclusions in a same diamond host, and evident trace element variations in some mineral inclusions show that some diamonds were formed by disequilibrium growth.

Keywords: diamond, inclusion in diamond, eclogite, trace element, subduction, Sino-Korea craton

INTRODUCTION

Lithospheric mantle beneath Archean cratons are composed of refractory peridotite and minor (~5%) eclogite. Geochemical properties of these mantle eclogite and then its genesis are important not only to the earth material circulation and diamond crystallization, but also to the formation of Archean lithophere. It is critical in evaluating models of Archean lithosphere formation to know the genesis of the eclogite inside. Despite extensive study, origin of mantle eclogite is still controversial. Contrasting models have been proposed, which include: (1) high-pressure igneous cumulates from basaltic melts within the upper mantle [1-4]; (2) metamorphic product, either by transformation of gabbro previously underplated onto the lower crust or by subduction of ancient oceanic crust [5-11]; (3) relics of the earth primary differentiation shortly after accretion [12].

Mineral inclusions in diamonds can be grouped into two major parageneses: peridotitic and eclogitic type. Usually peridotitic diamonds are much more abundant. Some special eclogite xenoliths, which contain diamonds with eclogitic mineral inclusions, have been found from the Siberian kimberlites. It was found that evident geochemical differences existed between the diamond inclusions and their counterparts in the host eclogite xenoliths, for both major and trace elements [10, 13]. Some inclusions in diamonds are more enriched, but some are more depleted in some incompatible elements, as comparing with the host eclogites. It strongly implies that partial melting and/or metasomatism may have substantially affected the compositions of mantle eclogite xenoliths in kimberlites, which makes discussion of geochemistry of mantle eclogite xenoliths be very complicate. Syngenetic mineral inclusions in diamonds have been effectively shielded from any chemical reaction or exchange with surroundings since the formation of host diamonds, and therefore can provide the first evidence of primary eclogite composition during diamond formation. Thus, comparing with eclogtie xenoliths, eclogitic mineral inclusions in diamonds are nearly original, and their compositions are more appropriate for discussing mantle eclogite genesis. Here, we give the first extensive geochemical study of eclogitic mineral inclusions from some Chineses diamonds. Major elements, and a large number of trace element (LILE, REE, HFSE) abundance in these inclusions were measured. Based on these results, genesis of mantle eclogite is discussed.

SAMPLE AND ANALYSIS

Studied diamonds with eclogitic mineral inclusions are from the "No.50" and "Shengli 1" kimberlite pipes in China. Geological settings of these kimberlite pipes may refer to Zhang et al. [14]. Diamonds from both pipes are dominated by peridotitic paragenesis. No eclogitic inclusion has ever been reported from the "No.50" pipe. Three eclogitic diamonds were found in this study, and inclusions were liberated by burning the host diamonds in air at 800°C for about eight hours. Totally 19 mineral inclusions were recovered, with sizes of 40 -250μm. Most of the inclusions are very fresh, and free from any alteration, except three in diamond S9. Details about inclusions and their host diamonds are listed in Table 1. Diamonds L32 and S12 contain pair inclusions of garnet and omphacite. Totally 12 inclusions were liberated from diamond L32 (4 omphacite, 7 garnet and 1 olivine). Coexisting of olivine, typical phase of peridotitic paragenesis, with eclogitic mineral inclusions in a same diamond host is an extremely rare, and its formation will be discussed later. All the inclusions generally display octahedral habits, a reflection of the confining diamond morphology.

Table1 Studied diamonds and their mineral inclusions

	Host diamonds		
Sample	L32	S9	S12
Pipe	No. 50	Shengli 1	Shengli 1
Size	~5 mm	~2.5 mm	~2.5 mm
Color	colorless	colorless	colorless
Shape	contact-twin	octahedron	octahedron
	Mineral inclusions		
Fe-rich garnet	7	5	1
Omphacite	4		1
Olivine	1		
Size (μm)	40~250	50~150	50~80
Freshness	fresh	2 fresh	fresh

Two omphacite inclusions, one from diamond L32 and the another from S12, show textures of partial melting, which are very similar to the "spongy" or "crinkled" texture observed around omphacite of eclogite xenoliths [8]. To the best knowledge of authors, this is the first report of such features from diamond inclusions. Major element compositions were measured by electron microprobe (JEOL-JXA8800), and analyses of trace elements were performed using a Cameca IMS-3F SIMS. Details may refer to other publications [15, 16]. Precision of analysis are believed to be within 10%.

MAJOR ELEMENT COMPOSITIONS

Extensive electron microprobe analysis shows that all fresh inclusions are homogeneous in

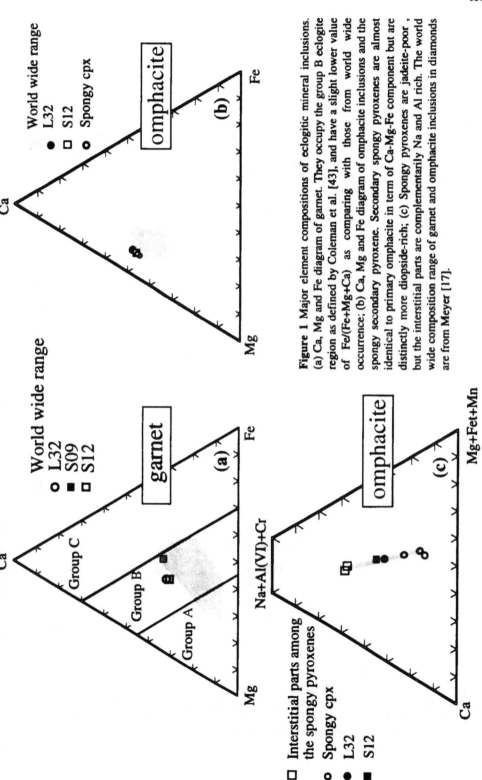

Figure 1 Major element compositions of eclogitic mineral inclusions. (a) Ca, Mg and Fe diagram of garnet. They occupy the group B eclogite region as defined by Coleman et al. [43], and have a slight lower value of Fe/(Fe+Mg+Ca) as comparing with those from world wide occurrence; (b) Ca, Mg and Fe diagram of omphacite inclusions and the spongy secondary pyroxene. Secondary spongy pyroxenes are almost identical to primary omphacite in term of Ca-Mg-Fe component but are distinctly more diopside-rich; (c) Spongy pyroxenes are jadeite-poor, but the interstitial parts are complementarily Na and Al rich. The world wide composition range of garnet and omphacite inclusions in diamonds are from Meyer [17].

major elements, and no intragrain variation was observed for garnet and pyroxene inclusions within a same diamond host. Major element compositions of these inclusions are depicted in Fig.1. Garnets are CaO and FeO rich, with range of CaO (11.85-12.60wt%), FeO (12.74-16.62 wt%) and MgO (7.95-11.01wt%). They occupy the group B eclogite region as defined by Coleman et al. [43], and have a slight lower value of Fe/(Fe+Mg+Ca) comparing with those of world wide occurrence (Fig.1a). Garnets in L32 and S12 have very similar major element compositions, but S9 is relatively rich in CaO, FeO, TiO_2 and Na_2O. Average TiO_2 and Na_2O contents in S9 garnets is 0.9 and 0.3wt%, respectively. Omphacite inclusions have a very limited composition range (~$Wo_{48}En_{42}Fs_{10}$). Na_2O and Al_2O_3 contents in S12 are slightly higher than that in L32 (Fig.1b,c), jadeite content is around 30-38mole%. Compositions of spongy pyroxene and its interstitial part, generated by partial melting of pristine omphacite , are listed in Table 2. The spongy pyroxenes are almost identical to primary pyroxenes in term of Ca-Mg-Fe component but are distinctly more diopsidic than unmelted counterparts, and interstitial parts among the spongy pyroxenes have high content of Na_2O and Al_2O_3 (Fig. 1b,c). These results are generally comparable with spongy pyroxenes in eclogite xenoliths [3], however, no evident increasing of K_2O content was identified contrast to that reported by Taylor and Neal [8] from eclogite xenoliths.

Table 2 Compositions of spongy pyroxene and its interstitial part

Sample*	SiO2	Al2O3	TiO2	Cr2O3	FeO	NiO	MnO	MgO	CaO	Na2O	K2O	Total
A	55.64	8.54	0.18	0.05	4.07	0.03	0.07	11.02	16.52	4.18	0.23	100.5
B	55.52	6.05	0.20	0.07	4.73	0.05	0.06	13.46	19.55	1.72	0.07	101.5
C	58.66	13.74	0.20	0.02	2.68	0.00	0.06	6.49	13.56	4.74	0.22	100.4

*A: Pristine omphacite inclusion (L32-02) before melting; B: Pyroxene in spongy part;
C: Interstitial parts among the spongy pyroxenes.

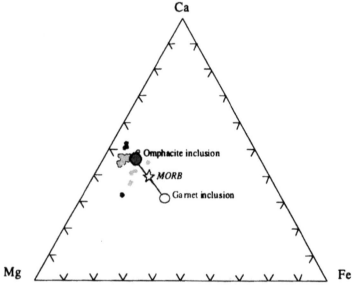

Figure 2 Average compositions of omphacite and garnet from Chinese diamonds. Bulk compositions of mantle eclogite included in diamonds are very close to that of MORB. Compositions of omphacite and garnet in diamondiferous eclogite xenoliths from the Udachnaya pipe in Siberia [18, 19] are also shown for comparison.

Average compositions of garnet and omphacite inclusions from Chinese diamonds are shown in Figure 2, comparing with those of diamond-bearing eclogite xenoliths from the Udachnaya kimberlite pipe in Siberia. It was found that both garnet and omphacite inclusions in the Chinese diamonds are more Ca and Fe richer than most diamond-bearing eclogite xenoliths. Bulk compositions of eclogite included in diamonds are very close to that of MORB. Olivine inclusion in diamond L32 has Mg/(Mg+Fe) value of 0.912. It is a little lower than normal olivine inclusions in diamonds, which are usually higher than 0.915.

TRACE ELEMENT GEOCHEMISTRY

Trace element concentrations of inclusions with suitable size were measured, and the results are listed in Table 3. Chondrite normalized rare earth element (REE) patterns of garnet inclusions are dsplayed in Fig.3. All the analyzed garnet inclusions are depleted in LREE, and

Table 3 Trace element contents of studied eclogitic inclusions (ppm)

Sample	S9(2)-1	S9(4)	S9(4)	S12-2	L32-03	L32-04	L32-06	L32-10	L32-11	L32-02	L32-02
Phase*	gt	gt	altered	gt	gt	gt	gt	omp	omp	omp	spongy
La	0.25	0.38	24.07	0.04	0.20	0.10	0.20	0.75	0.77	0.87	0.91
Ce	1.88	2.08	38.33	0.21	1.05	0.81	1.14	3.04	2.66	3.03	3.03
Pr	0.69	0.63	3.71	0.06	0.26	0.19	0.24	0.42	0.31	0.38	0.43
Nd	5.89	6.42	15.86	0.47	1.76	1.79	2.03	1.59	1.38	1.86	1.95
Sm	6.11	5.00	7.32	0.25	1.03	1.35	0.83	0.36	0.36	0.47	0.55
Eu	2.82	2.58	3.20	0.26	0.63	0.60	0.70	0.18	0.22	0.30	0.22
Gd	6.36	5.79	7.59	0.52	1.39	1.41	1.23	0.24	0.27	0.40	0.41
Tb	1.09	1.15	1.54	0.15	0.28	0.23	0.25	0.04	0.04	0.04	0.06
Dy	9.06	8.30	9.39	1.48	2.49	2.07	2.29	0.17	0.23	0.19	0.41
Ho	2.15	2.36	2.59	0.42	0.67	0.63	0.62	0.02	0.02	0.03	0.06
Er	6.20	5.70	5.60	1.30	1.93	1.85	1.77	0.05	0.05	0.05	0.07
Tm	0.95	0.88	0.84	0.22	0.30	0.30	0.25	0.01	0.01	0.01	0.01
Yb	5.97	5.07	4.32	1.42	1.86	1.87	1.66	0.03	0.02	0.03	0.04
Lu	0.78	0.69	0.86	0.22	0.35	0.25	0.29	0.003	0.002	0.004	0.003
Li	3.57	3.59	6.44	0.88	1.79	2.22	1.91	9.36	10.55	9.93	11.44
Be	1.73	2.21	2.18	2.16	2.02	2.53	8.72	1.36	1.06	1.09	0.98
B	0.15	0.15	4.05	0.10	0.09	0.07	0.21	0.16	0.37	0.38	0.35
Na	2680	2370	2225	821	772	746	784	25091	25823	27179	26687
K	23.3	5.8	182.6	0.9	1.8	1.2	53.2	1978	2055	2232	2421
Sc	65.4	58.2	54.4	79.2	77.8	71.3	64.3	10.1	14.9	17.2	15.6
Ti	5019	4724	4713	1564	1742	1721	1723	1014	933	966	1012
V	167	152	148	111	148	140	136	212	205	212	235
Co	91.4	67.6	64.6	67.4	70.5	66.9	66.5	33.9	35.3	38.3	42.9
Cu	18.8	13.7	12.8	10.5	7.1	6.1	9.1	23.8	26.3	31.0	26.8
Rb	12.3	8.6	9.8	7.2	7.0	6.0	4.6	2.0	2.2	2.9	4.8
Sr	14.8	14.1	20.6	1.1	3.5	3.5	4.5	233.6	213.3	231.3	215.4
Y	52.0	49.7	51.5	15.6	15.9	15.4	17.2	0.7	0.8	0.7	0.7
Zr	28.5	26.0	25.9	5.7	8.1	7.9	8.7	3.1	2.7	2.7	3.0
Nb	0.06	0.08	0.15	0.05	0.19	0.10	0.27	0.09	0.09	0.09	0.09
Cs	0.03	0.03	0.45	0.58	0.01	0.05	0.01	0.07	0.00	0.00	0.01
Ba	0.09	0.06	4.78	0.02	0.04	0.04	0.66	0.92	0.78	0.73	1.18
Hf	1.11	1.17	1.09	0.36	0.41	0.40	0.65	0.21	0.17	0.18	0.22

* gt: garnet; omp: omphacite; altered: altered part of garnet S9(4);
spongy: "spongy" part of omphacite L32-02.

have evident positive Eu anomaly. (La/Yb)n values are of 0.02~0.08, and Eu/Eu* of 1.32~2.12. Three garnet grains from L32 and two from S9 were analyzed, no difference in REE concentrations were detected among inclusions from same host, within uncertainty of measurement. Analyzed garnet inclusions also displayed a wide range of abundances (e.g., Lan=0.2~1.6; Ybn=9.0~37.5). Garnets from diamond L32 and S12 fall within the range of garnets from eclogite xenoliths of Siberia and southern Africa kimberlites [4, 18], but garnets in diamond S9 contain higher REE concentrations than the reported data (Fig.3). Very

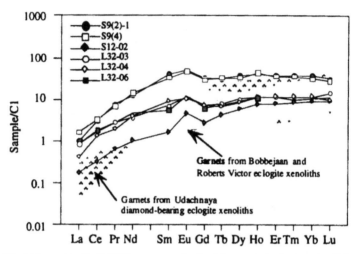

Figure 3 Chondrite normalized REE patterns of eclogitic garnet inclusions in Chinese diamonds. The most striking feature is the evident Eu positive anomalies in all the analyzed samples (Eu/Eu*: 1.32-2.12). Garnets from eclogite xenoliths of Africa and Siberia are also shown for comparison.

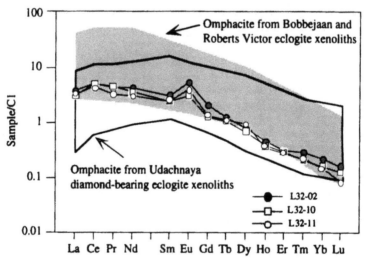

Figure 4 Chondrite normallized REE patterns of omphacite inclusions from diamond L32. No difference exists in these three separate grains. Eu positive anomaly is very clear (Eu/Eu* ~2.0). Omphacites from eclogite xenoliths of southern Africa and Siberia are also shown for comparison.

few REE data of eclogitic diamond inclusions are available for comparison. Garnets of this study have generally similar REE patterns with the ultradeep garnet inclusions from the Monastery diamonds, southern Africa [20].

Three omphacite inclusions from diamond L32 (L32-02; L32-10; L32-11) were analyzed for REE. No difference was detected among the three grains, within the uncertainty of measurement (Fig.4). They are enriched in LREE, and similar to garnet inclusions, positive Eu anomaly is also very evident. For SIMS measurement, molecular BaO has very close mass with Eu, and may affect the precision of Eu. But the observed Eu anomalies are not caused by the molecular BaO, because Ba contents in both garnet and omphacite inclusions are extremely low (usually <1ppm, see Table 3). (La/Yb)n value is about 19.05, and Eu/Eu* is around 1.96. Generally, it falls into the REE range of omphacites from eclogite xenoliths (Fig.4). No Eu anomaly was detected from the two Siberian omphacite inclusions [10, 13].

Concentrations of Be, B, Nb, Cs, Ba and Hf in garnet and omphacite inclusions are usually very low (<2 ppm). Na, K and Sr strongly partitioned into omphacite. Garnets have similar Y concentrations with the eclogitic garnet inclusions from Australian (Argyle and Ellendale pipe) diamonds, but Sr and Zr are much lower [21]. Range of Zr/Y is 0.36-0.54, much smaller than 2.7 of Australian garnet inclusions. Primitive mantle normalized [22] multiple element diagrams are shown in Fig.5, including large ion lithophile elements (LILE) (Rb, K, Ba, Sr),

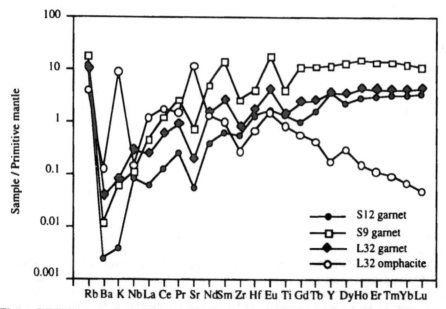

Figure 5 Primitive mantle normalized multiple element diagram of garnet and omphacite inclusions.

high field strength element (HFSE) (Nb, Ti, Zr, Hf) and REE. Average data of garnet and omphacite inclusions from a same host diamond are employed for drawing this figure. Generally, primitive mantle normalized concentrations of these elements increase gradually from left to right, but with evident enrichment in Rb and depletion in Sr and Zr and Hf, relative to the neighbor rare earth elements. Omphacite inclusion is strongly enriched in Rb, K and Sr, and also depleted in Zr and Hf. Bulk trace element compositions of mantle eclogite can be estimated from the coexisting garnet and omphacite inclusions, known modes of the two minerals. It is very difficult to know the modal proportions of minerals included in diamonds, but mode of garnet is usually within 30-70% for natural bimineral eclogite xeno-

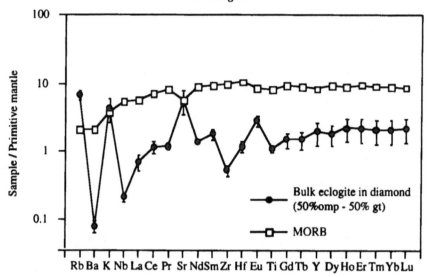

Figure 6 Primitive mantle normalized multiple element diagram of bulk eclogite occurred as inclusions in diamonds L32, with 50% garnet and 50% omphacite and 20% uncertainty of modal proportion. MORB is shown for comparison.

liths. Fig.6 shows the estimated bulk trace element compositions of mantle eclogite of 50wt% garnet - 50wt% omphacite from diamond L32, with 20% of uncertainty in modal proportion. Average mid-ocean ridge basalt (MORB) [23] is also displayed for comparison. MORB has a very smooth pattern, with the ratios increase gradually from left to right. Pattern of the estimated bulk eclogite (L32) does not change much with variation in mode of minerals. Except some elements, it is generally parallel to that of MORB, but with much lower values. Enrichment of LILE and depletion of HFSE are very clear, comparing with neighbor REE.

As pointed out previously, separate grains of a single phase within the same diamond host have very similar major and REE concentrations. However, some obvious differences also existed. Two garnet grains from diamond S9 exhibited distinct variation in K concentration. K content in garnet S9(2)-1 (23.3ppm) is about three times higher than that in garnet S9(4) (5.8ppm). In sample L32, three separate garnet grains were analyzed, among them L32-03 and L32-04 have very similar compositions for all the analyzed elements, but concentrations of Be, K and Ba in another garnet grain (L32-06) are about 4-40 times higher than the previous two grains (Table 3). All these variations in trace element concentrations can not be referred to analytical uncertainty or from contamination. Uncertainties of SIMS analysis for these elements are believed to be within ~10%. Ion beam was placed entirely within sample crystals, and surface contaminations are trifle. Trace elements in the "spongy" part of omphacite inclusion is also measured. Because ion beam diameter of ~30μm was employed for SIMS analysis, much bigger than the secondary pyroxene and interstitial materials, analytical results thus represent compositions of the whole spongy part. As listed in Table 3, concentrations for all analyzed trace elements in the spongy part are very close to that of the primary omphacite inclusion.

DISCUSSION

Equilibrium condition of the eclogitic mineral inclusions
Coexisting garnet and omphacite inclusions in a same diamond (sample L32) show no evident heterogeneity in major element compositions (Table 2), implying equilibrium between

the two phases has been generally achieved during the diamond growth in the eclogite environment. Equilibrium temperatures of these eclogitic inclusions in diamonds can be estimated using Ellis and Green's calibration of the Fe/Mg exchange between garnet and pyroxene[24]. Application of the Fe/Mg exchange geothermometer requires an independent estimate of pressure, which is not directly available from the mineral chemistry alone. In this study, equilibrium pressures were estimated by requiring a simultaneous solution with the ambient geothermal gradient at the time of kimberlite eruption (Fig.7). It has been definitely demonstrated that the lithospheric mantle beneath the Sino-Korea craton at the stage of kimberlite eruption (~460 Ma) is similar to that of the Kaapvaal craton, and has a cold geotherm very close to 40mw/m² [30]. According to this method, estimated equilibrium temperatures and pressures of the eclogitic mineral inclusions in diamonds are 1175-1240°C and 5.5-6.2GPa, corresponding to a depth of 160~190km in the upper mantle.

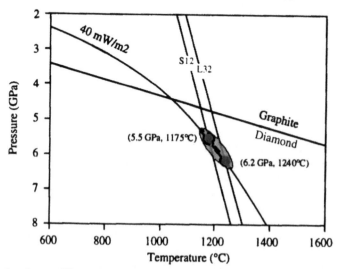

Figure 7 Estimation equilibrium conditions of eclogitic mineral inclusions diamonds. Lines labeled S12 and L32 are P-T relations determined by Mg and Fe exchange between garnet and omphacite.

Disequilibrium growth of diamond
Cathodoluminescence and infrared study of natural diamonds show that some diamonds have zoning structure, and repeated growing and dissolving for many times. It indicates that some diamond may be formed at a disequilibrium condition. Coexisting of olivine with eclogitic mineral inclusions in a same diamond host, like that of sample L32, is very unusual. It acts as the most compelling evidence for disequilibrium growth of the diamond, and implies that its formation may have taken a long period of time in the mantle. Similar "mixed" or "crossed" paragenesis diamonds have also been found in Africa, America and Australia [26-28]. In one diamond from the Monastery kimberlite in southern Africa, one olivine (mg# 94.5) coexists with an ultradeep eclogitic garnet, which contains significant amount of pyroxene solid solution [29].

Two possibilities are considerable for the formation of such "mixed" or "crossed" diamonds, transforming from peridotitic environment to eclogitic one or vice versa. Great deal of evidences, especially equilibrium temperatures of peridotitic inclusions in diamonds, show that peridotitic diamond crystallized at subsolidus condition. Thus, it is quite difficult for a peridotitic diamond to separate completely from its peridotite host rock and transmit to an eclogitic environment. Fig.8 shows the phase diagrams of peridotite and MORB (eclogite at high

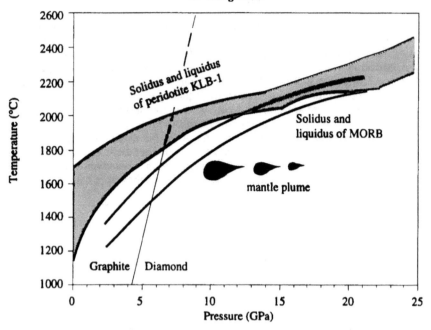

Figure 8 Formation of "mixed" or "crossed" diamond, which contains coexisting peridotitic and eclogitic inclusions. Melting temperature of MORB are distinctly lower than that of peridotite, and melting of eclogite within mantle plume can transmit eclogitic diamond into peridotitic environment easily in the diamond stable field. Phase diagram of KLB-1 is from Zhang and Herzberg [31] and the phase diagram of MORB from Yasuda et al. [32].

pressure). Liquidus temperature of MORB is apparently lower than the solidus of peridotite KLB-1 up to pressure of 12GPa. For example, at 4.0GPa and 1600°C, eclogite is entirely melted, but peridotite remains subsolidus. Considering an eclogite body within an mantle plume, we can find that diamond can be easily separated from the host eclogite and transmitted into peridotitic environment, because of high degree partial melting of eclogite and higher density of diamond. This process is illustrated in Fig.8. If this separation happened in diamond stable field, peridotitic mineral inclusions can be sealed in the eclogitic diamond during its re-growing in a peridotitic environment. Ultradeep garnet inclusions in diamonds, which contain substantial quantity of pyroxene solid solution, are believed to be transported from the transition zone by a "superplume" [33]. Coexisting of an ultradeep garnet with olivine inclusion (mg# 94.5) in a southern African diamond also supports this origin model. It is reasonable to image that diamond formed in eclogitic environment at the transition zone was transmitted into an environment like Archean lithospheric mantle, where the extremely refractory olivine was trapped during its subsequent growth. In addition to mantle plume, short-term thermal perturbations occurred repeatedly in the upper mantle, due to various reasons, for example, melt infiltrations. Such a local temperature increasing may trigger melting of eclogite but peridotite and thus creating a similar chance for transferring diamond from eclogitic environment to peridotitic one.

Another evidence for the disequilibrium growth of diamond is the heterogeneity of some trace elements in diamond inclusions. Multiple inclusions of a same phase within single diamond have very close major compositions, so do most trace elements (Table 3). However, some evident contrasts also exist, for example, concentrations of K, Be and Ba in garnets as mentioned previously. Similar observations have been reported by Griffin et al.

[21] for eclogitic inclusions in Australian diamonds, and by Shimizu and Sobolev [34] for peridotitic inclusions from Siberian diamonds. This implies that these diamonds grew in an open system, inclusions trapped at different stages of growth record changes in composition occurring in the host rock. Heterogeneously distributed K, Be and Ba are all extremely incompatible elements, and very sensitive to any mantle process, especially involving melt or fluid. General similarity of major elements and most trace elements among separate grains implies that the extent of diversion from equilibrium is not so significant, but the transition from eclogitic environment to peridotitic one like that occurred in diamond L32 is radical.

Formation of the spongy margin of omphacite inclusions
Omphacite in mantle eclogite xenoliths sometimes experienced various degrees of partial melting, which is reflected in the presence of spongy, symplectic textures [3, 8]. Spongy rim is composed of secondary pyroxenes and some interstitial materials. Taylor and Neal [8] proposed that the spongy margins was a reaction result of pristine pyroxene with metaso-matic melt, due to substantial quantities of K and Ba in the interstitial feldspar. To the best knowledge of author, no spongy margin of omphacite inclusion in diamond has never been reported. For the first time, this study confirms that such a partial melting can also occur in diamond inclusions. As shown in Table 2 and Figure 1, spongy pyroxenes are jadeite-poor , but the interstitial parts are complimentarily Na and Al rich, implying that the partial melting is an isochemcial process. It is impractical to image a chemical exchange with surroundings, because no crack was found in the diamonds. The most important evidence for an isochemical melting is trace element concentrations in both unmelted omphacite and the spongy part. As shown in Table 3, the unmelted pristine omphacite and its spongy margin have very similar concentrations for all the analyzed trace elements. Such a partial melting could be caused by temperature increase related to kimberlite eruption.

Genesis of mantle eclogite
Various opinions have been proposed as to the genesis of mantle eclogite. Among all others, major element compositional characteristics is the first feature need to be illustrated. As previously discussed, bulk composition of these eclogitic inclusions are broadly equivalent to basalt (Fig.2). Eclogitic mineral inclusions of this study were formed at pressure of 5.5 ~ 6.5GPa. At such a condition, basaltic melt cannot not be derived from fertile peridotite by anhydrous partial melting, according to high pressure melting experiments [30, 31, 35]. It has been demonstrated that the partial melts become increasingly MgO rich with increasing pressure. Basalt forms at pressure less than 3.0 GPa, picrite at about 3.0-4.0 GPa, and komatiitic magma at approximately 5.0-7.0GPa. Eclogitic inclusions in diamonds and dia-mond-bearing eclogite xenoliths, therefore, should be komatiitic in composition if they were formed by partial melting of fertile peridotite and not basaltic as is generally the case. The first liquidus phase of this MgO-rich melt, at this pressure range, should be olivine rather than garnet or clinopyroxene [36]. Hence, bulk major element compositions of mantle eclogite do not support the proposal that mantle eclogite to be high-pressure igneous cumulates. Age of mantle eclogite is another important factor for constraining formation of mantle eclogite. Pb-Pb and Re-Os isotope systems of Siberian diamond-bearing eclogite xenoliths in kimberlites show that these eclogites were formed in the Mid- to Late-Archean (2.7-2.9 Gyr) [11, 37]. This age is identical to the late-Archean Sm-Nd age of 2.7Ga observed in eclogite xenoliths from kimberlites of southern Africa [6, 38]. The Mid- to Late-Archean ages of eclogite do not favour an origin for these rocks as vestiges of the earth's primary differentiation in the Haden [12].

Positive Eu anomalies are very evident in all the analyzed mineral grains (Fig.3,4). It was also reported from eclogite xenoliths of African and Siberian kimberlites [18]. Eu^{2+} has a similar ion radius to Ca^{2+}, and can be enriched in plagioclase. It implies plagioclase fraction-ation is closely related to the formation of these mantle eclogite. Low density of plagioclase crystallizing in the basaltic magma chamber makes it very easy float and form plagioclase-rich cumulates. Thus, processes close to the earth surface are essentially necessary for the formation of mantle eclogite. There is no report of any possibility that melting and fractionation in the deeper mantle conditions can generate apparent Eu anomaly.

The petrological structure of oceanic lithosphere may vary markedly through geological time [39]. Mantle temperatures are believed to have been on average 200-300°C higher in the Archean than at present [40]. For this reason, the magmatic regimes at Archean spreading centers are likely to have differed considerably from those today. It is expected that higher mantle temperatures would produce more advanced degrees of partial melting, leading to copious generation of komatiitic magmas and a thickened oceanic crust. Komatiitic magmas may have fractionated near the crust-mantle interface by extensive crystallization of olivine, leading to the development of an oceanic crust composed mainly of tholeiitic basalts. The Archean oceanic lithosphere may have consisted of successive layers of basalt, olivine-rich cumulates and residual, refractory dunites [39]. Based on the previous discusses about genesis of mantle eclogite, it is reasonable to believe that mantle eclogite was formed by subduction of ancient oceanic crust.

However, very complex trace element properties, as shown in the multi-element diagram Fig.6, argue that mantle eclogite is not formed simply by subduction and recrystallization of ancient oceanic crust. Dehydration and partial melting during subduction may decrease the trace element abundances significantly, but can not cause LILE enrichment or HFSE depletion (Fig.6). Similar Zr, Hf depletion and Sr enrichment relative to neighbor REE were found from eclogitic inclusions in Udachnaya diamonds [10, 13]. Following two possible processes may have been involved in the formation of mantle eclogite. (1) Metasomatized by a carbonatite-like melt before recrystallization happened. Carbonatitic melts are usually enriched in LILE and LREE, but depleted in HFSE. Bulk mixing of small amount of such melts with basaltic component and subsequent recrystallization may form eclogite with the above trace element properties; (2) Kelemen et al. [41, 42] pointed out that mantle-magma interaction has an important role in controlling trace element behavior of mantle-origin rocks. For example, depletions of HFSE in arc basalts can be produced in melts by interaction with depleted mantle peridotite, because olivine and orthopyroxene have higher crystal/melt distribution coefficients for these elements than for other incompatible trace elements. They also found that this interaction may cause dissolving of Ca-pyroxene (cpx) and precipitating of orthopyroxene in mantle peridotite. Because cpx has high abundances of LILE, it is expected that this interaction may result in enrichment of these elements in the melt. Due to higher temperature of Archeam mantle, the subducted oceanic crust could be melted or at least high-degree partially melted during subduction. The melt generated there could also be basaltic in major element composition, and with similar Eu/Eu* ratio to original rocks. Interaction with Archean lithospheric mantle peridotite enriches LILE and depletes HFSE in the melt, but without causing evident change of its Eu. Crystallization of this melt in the deep mantle could be another mechanism for the formation of mantle eclogite.

As a summary for the discussion of mantle eclogite genesis, we believe that formation of eclogite in the diamond stability field was closely related to the subduction of ancient oceanic crust in the Archean. Some other processes like: (1) metasomatism by incompatible trace element rich melts, or (2) remelting and subsequent interactions with surrounding mantle peridotite may also have been involved.

Acknowledgements

Many thanks to Q. Miao for supplying the diamond sample L32. We are grateful to Prof. Z. Li and Dr. Z. Zhang for suggestions.

REFERENCES

1. J.R. Smyth, F.A. Caporuscio and T.C. McCormick. Mantle eclogite: evidence of igneous fractionation in the mantle. *Earth Planet. Sci. Lett.*, **93**, 133-141 (1989).
2. G.A. Snyder, E.A. Jerde, L.A. Taylor, A.N. Halliday, V.N. Sobolev and N.V. Sobolev. Nd and Sr isotopes from diamondiferious eclogites, Udachnaya kimberltie pipe, Yakutia, Siberia: evidence of differentiation in the early earth?. *Earth Planet. Sci. Lett.*, **118**, 91-100 (1993).
3. A.T. Fung and S.E. Haggerty. Petrography and mineral compositions of eclogites from the Koidu kimberlite complex, Sierra Leone. *Jour. Geophys. Res.*, **100**, 20451-20473 (1995).
4. F.A. Caporusico and J.R. Smyth,. Trace element crystal chemistry of mantle eclogites. *Contrib. Mineral. Petrol.*, **105**, 550-561 (1990).
5. H. Helmstaedt and R. Doig. Eclogite nodules from kimberlite pipes of the Colorado plateau - samples of Franciscan-type oceanic lithosphere. *Phys. Chem. Earth*, **9**, 95-111 (1975).
6. E. Jagoutz, J.B. Dawson, S. Hoernes, B. Spettel and H. Wake. Anorthositic oceanic crust in the Archean earth. *Lunar Planet. Sci.*, **15**, 395-396 (1984).
7. I.D. MacGregor and W.I. Manton. The Roberts Victor eclogites: ancient oceanic crust. *Jour. Geophys. Res.*, **91**, 14063-14079 (1986).
8. L.A. Taylor and C.R. Neal. Eclogites with oceanic crustal and mantle signatures from the Bellsbank kimberlite, South Africa, part I: mineralogy, petrography, and whole rock chemistry. *Journal of Geology*, **97**, 551-567 (1989).
9. C.R. Neal, L.A. Taylor, J.P. Davidson, P. Holden, A.N. Halliday, P.H. Nixon, J.B. Paces, R.N. Clayton and T.K. Mayeda. Eclogites with oceanic crustal and mantle signatures from the Bellsbank kimberlites, South Africa, Part 2: Sr, Nd, and O isotope geochemistry. *Earth Planet. Sci. Lett.*, **99**, 362-379 (1990).
10. T.R. Ireland, R.L. Rudnick and Z. Spetsius, Z. Trace elements in diamond inclusions from eclogites reveal link to Archean granites. *Earth Planet. Sci. Lett.*, **128**, 199-213 (1994).
11. D. Jacob, E. Jagoutz, D. Lowry, D. Mattey and Kudrjavtseva. Diamondiferous eclogites from Siberia: Remnants of Archean oceanic crust. *Geochim. Cosmochim. Acta*, **58**, 5191-5207 (1994).
12. D.L. Anderson. Hotspots, basalts, and the evolution of the mantle. *Science*, **213**, 82-89 (1981).
13. L.A. Taylor, G.A. Snyder, V.N. Sobolev, G. Crozaz and N.V. Sobolev. Trace element chemistry of eclogitic inclusions in diamond and comparisons with host eclogite, Mir kimberlite, Russia. *6th IKC Extended Abstract*, 635-627 (1995).
14. P. Zhang, S. Hu and G. Wan. A review of the geology of some kimberlites in China. In: *Kimberlites and Related Rocks, Vol.1, Their Composition, Occurrence, Origin and Emplacemen.* J. Ross (Ed.). pp.329-400 (1989).
15. H. Yurimoto, A. Yamashita, N. Nishida and S. Sueno. Quantitative SIMS analysis of GSJ rock reference samples. *Geochemical Journal*, **23**, 215-236 (1989).
16. W. Wang and S. Yurimoto. Analyis of rare earth elements in garnet by SIMS. *Annual Report, Institute of Geoscience, University of Tsukuba*, **19**, 87-91 (1994).
17. H.O.A. Meyer. Inclusions in diaomond. In: *Mantle Xenoliths.* P.H. Nixon (Ed.), pp. 501-523 (1987).
18. E.A. Jerde, L.A. Taylor, G. Crozaz, N.V. Sobolev and V.N. Sobolev. Diamondiferous eclogites from Yakutia, Siberia: evidence for a diversity of protoliths. *Contrib. Mineral. Petrol.*, **114**, 189-202 (1993).
19. V.N. Sobolev, L.A. Taylor, G.A. Snyder and N.V. Sobolev. Diamondiferous eclogites from the Udachnaya kimberlite pie, Yakutia. *Inter. Geo. Rev.*, **36**, 42-64 (1994).
20. R.O. Moore, J.J. Gurney, W.L. Griffin and N. Shimizu. Ultra-high pressure garnet inclusions in Monastery diamonds: trace element abundance patterns and conditions of origin. *Eur. J. Mineral.*, **3**, 213-230 (1991).

21. W.L. Griffin, A.L. Jaques, S.H. Sie, C.G. Ryan, D.R. Cousens and G.F. Suter. Conditions of diamond growth: a proton microprobe study of inclusions in West Australian diamonds. *Contrib. Mineral. Petrol.*, **99**, 143-158 (1988).
22. W.F. MacDonough and S.-s. Sun. The composition of the earth. *Chemical Geology*, **120**, 223-253 (1995).
23. A.W. Hofmann. Chemical differentiation of the earth: the relationship between mantle, continental crust, and oceanic crust. *Earth Planet. Sci. Lett.*, **90**, 297-314 (1988).
24. D.J. Ellis and D.H. Green. An experimental study of the effect of Ca upon garnet-clinopyroxene Fe-Mg exchange equilibria. *Contrib. Mineral. Petrol.*, **71**, 13-22 (1979).
25. W. Wang, E. Takahashi and S. Sueno. Composition of the lithospheric mantle beneath the Sino-Korea craton. *6th IKC Extended Abstract*, 663-665 (1995).
26. M. Prinz, D.V. Manson, P.F. Hlava and K. Kiel. Inclusions in diamonds: garnet lherzolite and eclogite assemblages. *Phys. Chem. Earth.*, **9**, 797-815 (1975).
27. A.E. Hall and C.B. Smith. Lamporite diamonds - are they different? In: *Kimberlite occurrence and orgin: A Basis for Conceptual Models in Exploration.* J.E. Glover and P.G. Harris (Ed.). pp. 167-212 (1984).
28. M.L. Otter, and J.J. Gurney. Mineral inclusions in diamond from the Sloan diatremes, Colorado-Wyoming State Line kimberlite district, North America. In: *Kimberlites and related rocks.* J. Ross (Ed.). 2, 1042-1053 (1989).
29. R.O. Moore and J.J. Gurney. Mineral inclusions in diamond from the Monstery kimberlite, South Africa. In: *Kimberlites and related rocks.* J. Ross (Ed.). 2, 1029-1041 (1989).
30. E. Takahashi. Melting of a dry peridotite KLB-1 up to 14 GPa: implications on the origin of peridotitc upper mantle. *Jour. Geophys. Res.*, **91**, 9367-9382 (1986).
31. J. Zhang and C. Herzberg. Melting experiments on an anhydrous peridotite KLB-1 from 5.0 to 22.5 GPa. *Jour. Geophys. Res.*, **99**, 17729-17742 (1994).
32. A. Yasuda, T. Fujii and K. Kurita. Melting phase relations of an anhydrous mid-ocean ridge basalt from 3 to 20 GPa: Implications for the behaviour of subducted oceanic crust in the mantle. *Jour. Geophys. Res.*, **99**, 9401-9414 (1994).
33. S.E. Haggerty. Superkimberlite: A geodynamic diamond window to the Earth's core. *Earth Planet. Sci. Lett.*, **122**, 57-69 (1994).
34. N. Shimuzu and N.V. Sobolev. Young peridotitic diamonds from the Mir kimberlite pipe. *Nature*, **375**, 394-397 (1995).
35. E. Takahashi and C.M. Scarfe. Melting of peridotite to 14 GPa and the genesis of komatiite. *Nature*, **315**, 566-568 (1985).
36. K. Wei, R.G. Tronnes and C.M. Scarfe. Phase relations of Aluminum-Undepleted and Aluminum-Depalted komatiites at pressures of 4-12 GPa. *Jour. Geophys. Res.*, **95**, 15817-15827 (1990).
37. D.G. Pearson, G.A. Snyder, S.B. Shirey, L.A. Taylor, R.W. Carlson and N.V. Sobolev. Archean Re-Os age for Siberian eclogites and constraints on Archean tectonics. *Nature*, **374**, 711-713 (1995).
38. W.I. Manton and M. Tatsumoto. Some Pb and Sr isotopic measurements on eclogites from the Roberts Victor mine, South Africa. *Earth Planet. Sci. Lett.*, **10**, 217-226 (1971).
39. A.E. Ringwood. Slab-mantle interactions 3. Petrogenesis of intraplate magmas and structure of the upper mantle. *Chemical Geology*, **82**, 187-207 (1990).
40. M.J. Bickle. Implications of melting for stabilisation of the lithosphere and heat loss in the Archean. *Earth Planet. Sci. Lett.*, **80**, 314-324 (1986).
41. P.B. Kelemen, K.T.M. Johnson, R.J. Kinzler and A.J. Irving. High-field-strength element depletions in arc basalts due to mantle-magma interaction. *Nature*, **345**, 521-524 (1990).
42. P.B. Kelemen, H.J.B. Dick and J.E. Quick. Formation of harzburgite by pervasive melt/rock reaction in the upper mantle. *Nature*, **358**, 635-641 (1992).
43. R.G. Coleman, D.E. Lee, L.B. Beatty and W.W. Brannock. Eclogites and eclogites: their differences and similarities. *Geol. Soc. Am. Bull.*, **76**, 483-508 (1965).

Proc. 30thInt'l. Geol. Congr. , Part15, pp. 199 – 213
Li et al. (Eds)

Genesis and Two Magmatic Evolution Trends of the Shuiquangou weakly Alkaline Complex, Northern Hebei Province, North China

ZHANG ZHAOCHONG and LI ZHAONAI

Institute of Geology, Chinese Academy of Geological Sciences, Beijing, 100037, *P. R. C.*

Abstract

The Shuiquangou complex is elongated in an east-west direction, which accords with the direction of deep-fault. It is characterized by enrichment in alkali and Fe and poor in Ca. The petrochemical characteristics indicate that the complex is a potassic to high-potassic complex consisting of cala-alkaline, weakly alkaline and alkaline rocks. Their petrochemical diagrams, REE and trace elements patterns show that the magma has two evolutional trends: acid (pyroxene diorite → amphibole monzonite → quartz alkali-feldspar syenite → alkali-feldspar granite) and alkaline (syenite → aegirine-augite syenite → alkali-feldspar syenite). The relatively high concentration of Ba, K, Sr and Hf with relatively low concentrations of Rb, Th, Nb, P and Ti implies that it should orignate from the mixed source region of mantle and crust. The calculation based on the ^{87}Sr/^{86}Sr ratio gives a mixing proportion: 62% upper mantle materials and 38% crustal materials, which suggests that the primary magma derived from the transition of the upper mantle and lower crust.

Keywords: weakly alkaline complex, genesis, two magmatic evolutional trends, mixed source materials, Northern Hebei Province

INTRODUCTION

Since tens gold deposits with gold reservers totalling about 100 metric tons were discovered in the Shuiquangou complex in recent years, such as Dongping, Hougou and Zhongshangou deposits, the importance of the relationship between the Shuiquangou Complex and the gold deposits has been gradually recognized [31]. The large area, precipitous terrain and dense vegetation in the outcroup region of the complex hinder the detail study for it to some extent. Many researchers focused their research on the ore-bearing country rocks and gold deposits [10,17,22]. The genesis of the complex is still being argued. The main views for it are: magma differentiation after partial melting of upper mantle or lower crust rocks [20,21,23], metmorphic agpaite volcanic rock series [24], remelting, metasomatism and migmatization [10,11,17,22], which is the most popular view. There is no common view about the genesis of the complex with mainly intermediate weakly alkaline rocks like the Shuiquangou Complex in the world; assimilation and contamination [12]; metasomatism and partial melting of mantle rocks [3]; mixing of magma [4]; liquid immiscibility [1] and fractional crystallization [9]. Therefore, it is very important to study the genesis of Shuiquangou Complex for both solving the petrogenesis of the inntermediate weakly alkaline rocks and prospecting for gold deposits related to the alkaline rocks. In this paper, we try to provide a clue for solving the origin of intermediate weakly alkaline magma via discussion on the origin of the Shuiquangou weakly alkaline complex.

REGIONAL GEOLOGY

The complex is tectonically located in the middle part of north margin of China-Korea plate, the south of the Shangyi—Chicheng Deep Fault in an approximately east-west direction which marks the boundary between the Yanshan subsidence belt and the Inner Monglia axis (Fig. 1). This area is a relatively active block. The regional stratigraphic sequence is an Archean amphibole facies of metamorphic rocks of the Sanggan Group, overlain by Proterozic marine sedimentary systems of the Changcheng and Jixian formations, and then by Jurassic volcanic-sedimentary rocks. There are three cycles of the main tectono-magma activies: Archean-Proterozoic, Hercynian-Indosinian and Yanshanian. Most of the intrusions distributed along the south side of Shangyi — Chicheng Deep Fault; The weakly alkaline intrusions and ultrabasic intrusions outcrop in EW trend, and some small acid intrusions mainly outcrop in NE trend, which shows that different intrusions are controlled by the different-striking faults.

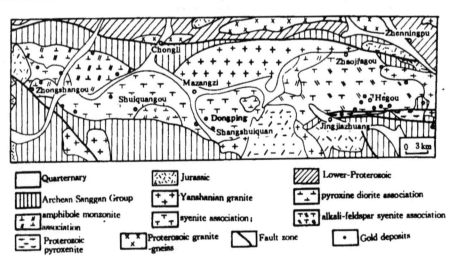

Fig. 1.   Geological sketch map of the Shuiquangou complex, northern Hebei Province

GEOLOGY OF THE COMPLEX

The Shuiquangou Complex, controlled by the Shangyi—Chicheng Fault, out-crops in an EW-trend, with an area of 400 km², about 56 km in length, 6—8 km in width. The periphery of the complex dip outside, but the dip angle at north side is about 70°—80° and the dip angle at the south side is gentle, about 40°—50°. Its country rocks are diopside-amphibole leptynite, gnesis, amphibo-lite and pyroxene granulite of the Jiangouhe Formation of Sanggan Group; the west and middle parts of the complex are covered by Upper-Jurassic Zhangjiak-ou and Baiqi formations respectively. Its perphery is also intruded by several Yanshanian granite plutons (Fig. 1).

The boundary between the complex and its country rocks was considered to be obscure before, but we found the clear boundary between them, which shows the complex is not resulted from either metasomatism or migmatization.

The enclaves are well-developed in the complex, especially near the contact zone. The enclaves consist mainly of the xenolithes of the country rocks ,such as gnesis, marble and amphibolite, but cognate enclaves have not been found yet. The compositions of the xenolithes were transformed to some extent that are now rich in darker minerals and poor in K-feldspar in comparision to the o-riginal rocks.

The age of the complex is still being debated. About 50 isotope ages(K-Ar, Ar-Ar and Rb-Sr isochron) define a very wide range of 120 Ma to 309 Ma, with a crest age value of 190 Ma to 210 Ma, which implies that magma activity began in Hercynian period and became intense in Indosian period (Triassic). The young ages could be consided to be altered time. This feature accords with the range of the ages of the regional weakly alkaline complexes [13,25].

On the basis of intrusive sequences, mineral paragenetic associations, and the characterestics of the accessory minerals, we divided the complex into four asso-ciations: a pyroxene diorite association, an amphibole monzonite association, a syenite association and an alkali-feldspar syenite association. The distribution of the various associations is seen in Fig. 1. Each association consists of several types of rocks. Most of them are gradual change relationship, and only a few are of mutation among them. There is no circular-zoning as the general com-plex.

GEOCHEMISTRY

Major Element Geochemistry

The characteristics of the major element have been studied and reported on in detail by Zhang et al. (1996) [32], and only a brief of summary of the results is provided here. The SiO_2 contents of the complex range from 57.37% to 74.76%, but most of them range from 61% to 66%. The complex is characterized by high alkali contents with the alkalinity of 6.77 to 14.00 and the Rittman Index (σ) of 2.85 to 8.64. Pyroxene diorite is the lowest alkalinity rock belonging to cala-alkaline series. The alkali-feldspar granite is in high alkalinity but low σ value of 2.89 to 3.19, averaging 3.07, belonging to calc-alkaline series. Other rock types belong to weakly alkaline and alkaline series, which are dominated parts in the complex. Most types of rocks with Hy norm and without Wo norm show the relatively high Fe contents and the low Ca contents. The Al values ($Al_2O_3/CaO+Na_2O+K_2O$) of the various rocks range from 0.75 to 1.20, but most Al values are less than 1.0 and close 1.0.

There are many definations about the alkaline igneous rocks. When discussing on the relationship between the precious metal deposits and the Cordilleran alkaline igneous rocks, Mutschler (1991) also studied alkaline igneous rocks in detail. After synthesizing the Mcdonald (1968) and Irvine's (1971) classification for alkaline igneous rocks, he gave his classification: $Na_2O+K_2O(\%)>0.3728SiO_2(\%)-14.25$ (Fig. 2), is alkaline; $Na_2O+K_2O(mol)\geqslant Al_2O_3$, is peralkaline, and the feldspathoid can not be used as the standard of the classification. We use the Mutschler's method here because of the close relationship between the Shuiquangou Complex and the gold deposits. Fig. 2 shows that pyroxene diorite, quartz alkali-feldspar syenite and alkali-feldspar granite belong to the sub-alkaline rocks; amphibole monzonite, alkali-feldspar syenite, syenite and aegirine-augite syenite belong to the alkaline rocks. The moles of Na_2O+K_2O of all rocks are less than those of Al_2O_3, so all of them don't belong to the peralikaline rocks. Diagram of K_2O vs. Na_2O (Fig. 3) shows that most rocks are plotted in potassic field; few of them such as alkali-feldspar syenite are plotted in high-potassic field; quartz alkali-feldspar syenite is plotted near the boundary between high-potassic and potassic field. Consquently, the Shuiquangou complex is a potassic to high-potassic complex consisting of calc-alkaline, weakly alkaline and alkaline rocks.

The Q—Ne—Kp phase diagram (Fig. 4) shows that there are two kinds of evolutional trends, one is toward the acid (point M_1); the other is toward alkaline (point M_3), but no trend toward point M_2. This character is consistent with the conclusion that the complex belongs to the potassic series, and also with the mineral paragenetic association that there is no feldspathoid association, but two kinds of alkali feldspars or quartz+alkali feldspar. Pyroxene diorite, amphibole monzonite, quartz alkali-feldspar syenite and alkali-feldspar granite show acid differentiation trend; syenite, aegirine-augite syenite and alkali-feldspar syenite

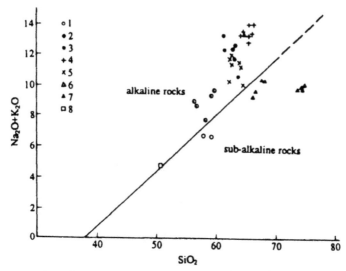

Fig. 2　Diagram of Na$_2$O+K$_2$O vs. SiO$_2$. 1=pyroxene diorite; 2= amphibole monzonite (including monzonite); 3=aegrine-augite syenite; 4=alkali-feldspar syenite; 5= syenite; 6=quartz alkali-feldspar syenite; 7= alkali-feldspar granite; 8=plagioamphibolite

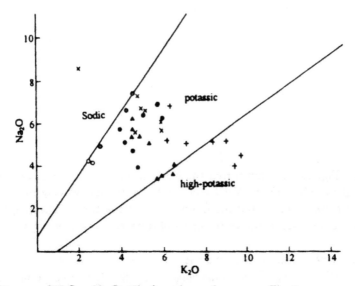

Fig. 3　Diagram of K$_2$O vs Na$_2$O. The legends are the same as Fig. 2 show alkaline differentiation trend.

The diagram of differentiation index (DI) vs. major oxides (Fig. 5) clearly

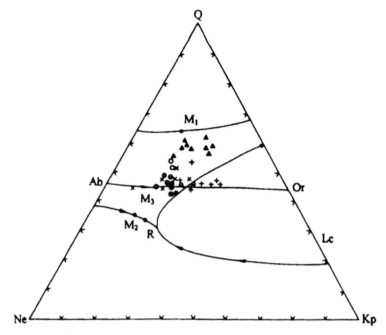

Fig. 4   Diagram of Q—Ne—Kp. The legends are the same as Fig. 2

shows the two lines of petrochemical variation. The end of one variation line is
alkali-feldspar granite, and the end of the other line is alkali-feldspar syenite.
This character indicates the two evolutional trends; acid and alkaline. The acid
evolutional trend is from pyroxene diorite to amphibole monzonite, to quartz alkali-feldspar syenite, and then to alkali-feldspar granite; The alkaline evolutional trend is from syenite to aegirine-augite syenite, and then to alkali-feldspar
syenite. The variation of these two evolutional lines are overall similar. Their
main difference is that Na_2O and K_2O contents increase gradually with the DI
increase in the acid evolutional line while the Na_2O decrease rapidly and the K_2O
increase rapidly with the DI increase in the alkaline evolutional line. This diagram (Fig. 5) also shows that the FeO, MgO and CaO decrease rapidly at low
DI value, while FeO, MgO and CaO change unobviously at high DI value.
These features show that the fractional crystallization of ferromagesian minerals
mainly take place in the early period of differentiation, not in the late period.

Rare earth element and trace element Geochemistry
The REE patterns of the Shuiquangou Complex have two groups of nearly parallel curves (Fig. 6), all of which belong to the low to middle enrichment in
LREE. This character is different from those of the Archean metamorphic rock
of Sanggan Group, which is characterited by the even type of REE pattern.
Another character of the REE pattern is absence of negative anomaly at Eu
which indicates no fractional crystallization of plagioclase or not the products of
partial melting of upper crust.

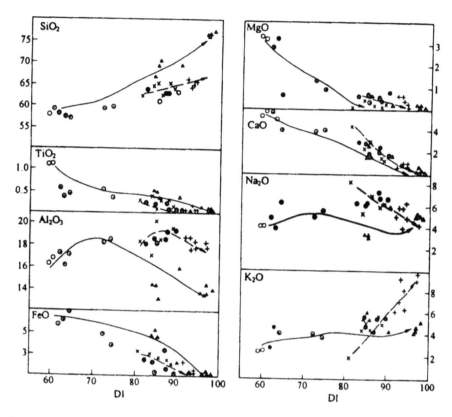

Fig. 5 Diagram of DI vs major oxides. The solid line arrow represents the acid evolutional trend, and the dotted line arrow represents alkaline evolutional trend. The other legends are the same as Fig. 2

The diagram of REE vs Ti, Zr and P shows that the total REE content is positive correlation with Ti, Zr and P, especially with Ti [28], which reflects that the total REE content is mainly enrichment in the accessory minerals, such as titanite, apatite and zircon. The fractional crystallization of these accessory minerals is the main reason leading the fractionation of rare earth elements.

The spider-web diagrams of trace elements also have two groups of nearly parallel curves (Fig. 7). They exhibt the enrichment in LILEs as well as depletion in HFSEs, and the positive anomalies of Ba, K, Sr and Hf as well as the negative anomalies of Rb, Th, Nb, P and Ti, which shows the mixed characteristics of mantle and crust sources, and different from that of the Archean metamorphic rocks.

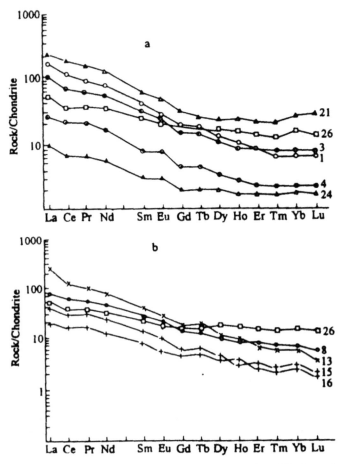

Fig. 6. REE patterns of various rocks of the Shuiquangou complex. The legends are the same as Fig. 2. a—acid evolutional trend; b—alkaline evolutional trend

Isotope Geochemistry

The $\delta^{18}O$‰ value of pyroxene diorite is 2.9, being lowest among the different rocks of the Shuiquangou complex. Those of amphibole monzonite are 5.88 to 7.9, averaging 7.03. Those of syenite are 4.1 to 8.13, averaging 6.13. Those of syenite to alkali-feldspar syenite in the alkali-feldspar syenite association in Hougou area are 7.1 to 9.4, averaging 8.17. Those of quartz alkali-feldspar syenite and alkali-feldspar granite are 10.32 and 7.6 respectively. Therefore, except for few samples with those higher than 10 or lower than 5.5, the $\delta^{18}O$‰ values of the most samples are between 5.5 and 10.0, that is, belonging to the normal granitiod. It implies that the magmas of the Shuiquangou complex mainly originated from the upper mantle, but contaminatd by crustal materials.

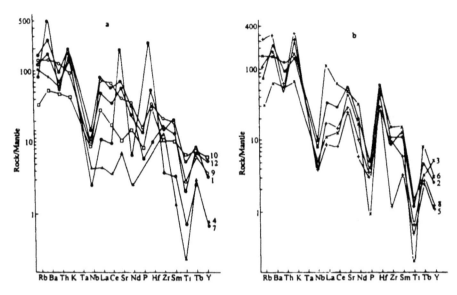

Fig. 7   Spider-web diagram of trace elements of the Shuiquangou complex. The legends are the same as Fig. 2. a—acid evolutional trend; b—alkaline evolutional trend

The lead isotope compositions of the Shuiquangou complex are: $^{206}Pb/^{204}Pb = 16.4565 - 17.3904$, $^{207}Pb/^{204}Pb = 15.2701 - 15.4723$, $^{208}Pb/^{204}Pb = 36.5321 - 37.3931$. The lead isotope values of the ampibole monzonite are slightly lower than those of the aegirine-pyroxene syenite and alkali-feldspar syenite. The lead isotope compositions of the Archean metamorphic rocks are: $^{206}Pb/^{204}Pb = 14.32 - 14.70$, $^{207}Pb/^{204}Pb = 14.77 - 14.94$, $^{208}Pb/^{204}Pb = 34.0925 - 34.3314$. Those values are much lower than those of the Shuiquangou complex. Those characteristics suggest that the relatively lower lead isotope values of amphibole monzonite are probably related to the contamination of the Archean strata, and that the Shuiquangou complex didn't originate from the Archean metamorphic rocks of the Sanggan Group.

The six projected points are plotted on the both sides of the upper-mantle growth line on the diagram of the lead structural model of Zartman and Doe (1981) [26]. Amphibole monzonite, syenite and aegirine-augite syenite are located between the two growth lines of upper-mantle and lower-crust, and alkali-feldspar syenite is located between the two growth lines of upper-mantle and orogen. The orogen lead represents the mixed lead of sedimentary materials and magmas melted from the deleted mantle in the mantle wedge region. Therefore, the material source of the Shuiquangou complex is not unique source region, but possibly the mixed sources of upper-mantle and lower-crust.

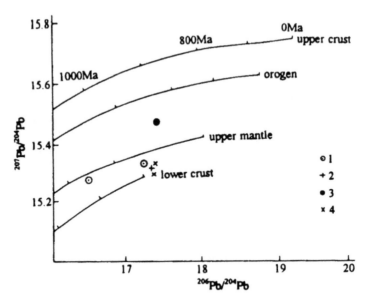

Fig. 8. Lead-isotope composition of the Shuiquangou complex plotted on the lead structural model of Zartman and Doe (1981). 1—amphibole monzonite; 2—alkali-feldspar
syenite; 3—aegirine-augite syenite; 4—syenite

Five samples of the Shuiquangou complex were employed to be measured for
Rb-Sr isochron age. A Rb-Sr isochron age of 309 ± 14Ma and an inital value of
$^{87}Sr/^{86}Sr$ of 0.70618 ± 0.00014 with the related coefficient of 0.9969 were
obtained. It can be obviously confirmed that the magmatic source belongs to the
type of the mixed source of mantle and crust (MC type).

The mixed strontium model of two members given by Faure (1986) is used as
follows[6];

$$I'_{Sr,m} = \frac{I'_{Sr,A}X_A f_A + I'_{Sr,B}X_B(1-f_A)}{X_A f_A + X_B(1-f_A)} \tag{1}$$

Whereas $I'_{Sr,A}$ and X_A is the strontium isotope value and strontium content of
crust at t time respectively; $I'_{Sr,B}$ and X_B is the Sr isotope value and Sr content of
mantle respectively; $I'_{Sr,m}$ is the mixed Sr isotope value; $f_A = \dfrac{A}{A+B}$, respre-
senting the percentage of crustal materials. We can convert Eq. (1) to have the
following expression;

$$f_A = \frac{(I'_{Sr,m} - I'_{Sr,B})X_B}{(I'_{Sr,A} - I'_{Sr,m})X_A + (I'_{Sr,m} - I'_{Sr,B})X_B} = \frac{1}{1+P} \tag{2}$$

$$Whereas\ P = \frac{I'_{Sr,A} - I'_{Sr,m}}{I'_{Sr,m} - I'_{Sr,B}} \frac{XA}{XB} \tag{3}$$

The inital $^{87}Sr/^{86}Sr$ value of the metamorphic rocks of the Sanggan Group was measured by Gao et al. (1988) [7]. The $I'_{Sr,A}$ value at 300 Ma can be calculated by the approximate expression given by Faure (1972) [5]:

$$(^{87}Sr/^{86}Sr)_t \approx (^{87}Sr/^{86}Sr)_0 + (^{87}Rb/^{86}Sr)\lambda t \tag{4}$$

From the Eq. (4), $I'_{Sr,A} = 0.7094$, and its Sr content is 490×10^{-6}. The Sr isotope value of primary mantle suggested by Faure (1986) is 0.704, and its Sr content is 450×10^{-6}[6]. The f_A is calculated to be 38% from Eq. (2) and (3). That is, the source materials consist of 38 percent crustal materials and 62 percent mantle materials.

Ascent Velocity and Emplacement Depth

Under the constant pressure gradient (∇P), the magmatic ascent velocity can be calculated from the following expression when magmas ascend along the vertical fissure with a constant width [8]:

$$U_m = \nabla P \cdot \Delta x^2 / 12\mu \tag{5}$$

whereas U_m—ascent velocity, Δx—the width of fissure, μ—magmatic viscosity. When the pressure gradient results from density differenc between magmas and their country rocks, the Eq. (5) can be converted as the following:

$$U_m = (\rho_s - \rho_f)\Delta x^2 / 12\mu \tag{6}$$

whereas ρ_s and ρ_f respresent the density of magmas and country rocks respectivly. The Shuiquangou complex is controlled by the Shangyi—Chongli—Chichen deep fault, which width, 10 metres, can be approaximately taken grant for the fissure width of magma ascent. The density of deep crust (country rocks) is about 2.75×10^3 Kg/m³. The magmatic density and viscosity calculated from the expressions by Botting et al. (1972) [2] and Shaw (1972) [16] is 2.37×10^3 Kg/m³ and 4.55×10^7 Pa · S respectively according to the compositions of major magmas. Thus, the magma ascent velocity is estimated to be 7×10^{-5} m/s, that is ascending about 311 m each year. Compared with that of the granitic magmas from other area, the magma ascent velocity of the Shuiquangou complex is much slower. Such slow ascent velocity is favourable for the contamination of magmas with their country rocks.

The geological method is the most reliable method for the estimation of magmatic emplacement depth. There are many isolated xenolithes of marbles, appearing the same occurances, in the Shuiquangou complex, which shows the roof pendant of the complex. These marbles locally occur in the unconfirmity with the gneiss of the Sanggan Group. Therefore, the marbles in the complex are not interbeds of the Sanggan Group, but belong to the Proterozoic Changchen Formation. As the illustration above, the main implacement age of the Shuiquangou complex is Indosian epoch. Thus, the thickness of the overcovered strata of the intrusion, from Preterozoic Changchen Formation to Permian system, should represent the emplacement depth of the Shuiquangou complex (2—3 km).

DISCUSSIONS

The petrochemical, REE and trace element characteristics of the Shuiquangou complex have revealed the two magmatic evolutional trends, and the REE, trace elements and isotope (O, Pb and Sr) characteristics show that the magmatic source of the Shuiquangou complex is the mixed source of crust and mantle, consisting of about 38% crustal materials and 62% mantle materials according to the strontium isotope compositions. So the key problems for the origin of the Shuiquangou complex are: the main way of the magma mixing and the key factors controlling the magmatic evolution.

The most basic rock type of the Shuiquangou complex is pyroxene diorite, which contains SiO_2 content of 58—59%, analogous to the andesitic magmatic composition, but absence of basic rock in the complex. If the primary magma originated from the upper-mantle, a few intrusions with basic compositions should have been formed. Therefore, the complex didn't originate from the upper-mantle directly. The characteristics of the garnets in the Shuiquangou complex indicate that they are formed by contamination [30], but it is impossible for magmas to contaminate so many crustal materials (38%). When 38% crustal materials were put in the magmas, only several percent crustal materials would be contaminated, and a large amount of materials would have remained in the magmas. It is not in accord with the fact. So, the contamination, locally occuring, is not the major reason leading the mixing of about 38% crustal materials. The liquid—liquid mixing, a mixing of the magmas from mantle with the magmas from crust, is not only absence of any geological evidences (e. g. , without disequilibrium mineral ssociations), but also doesn't supported by the REE, trace elements and isotope characteristics. The solid—solid mixing is generally considered to be in the mantle wedge where the oceanic crustal materials mix with the mantle materials because of the subduction of oceanic crust. The researched region is located within the continental plate far away from the subduction zone at Hercynian period [15]. Consequencely, this model of solid—solid mixing is not suitable for the researched region. Another model of solid—solid mixing, the mixing of magma source region materials, is more possible. Such source region should be the location between the top of upper mantle and bottom of lower-crust where is the transition of the upper mantle and lower crust. In this location, the upper mantle materials contain the lower crustal materials, while the lower crustal materials also contain the upper mantle materials, which has been confirmed by geophysical information. The original magma melted from such source region should be the intermediate magma. It homogenerously mixed at high temperature, but when the temperature decreased, the liquid immiscility of the homogenerous mixed magmas formed two parental magmas, which formed the two evolutional series by the differentiation at shallow-level magma chamber. The contamination occured at the sides of shallow-level magma chamber and ascending process because of the slow ascent velocity. The magmas finally emplaced at about 2 — 3 km depth of surface downward. This infered process accords with the geological and geochemical characteristics.

CONCLUSIONS

The Shuiquangou intrusion is a potassic to high-potassic complex consisting of calc-alkaline, weakly alkaline and alkaline rocks.

The magma has two evolutional trends. One trend is acid: pyroxene diorite→ amphibole monzonite→ quartz alkali-feldspar syenite→ alkali-feldspar granite, and the other trend is alkaline: syenite → aegirine-augite syenite → alkali-feldspar syenite.

The primary magma originated from the transition of the upper mantle and lower crust, consisting of about 62% mantle materials and 38% lower crustal materials. The depth of the magma emplacement is about 2 — 3 km. The complicate associations of the Shuiquangou complex were resulted from partial melting of the mixed materials of crust and mantle, liquid immiscibility, differentiation (including fractional crystallization of the ferromagnesian minerals at early time) and contamination at shallow-level magma chamber and ascending process.

Acknowledgments

The project supported by National Natural Science Fundation of China is gratefully acknowledged. We appreciate the managements of the Dongping and Hougou gold mines for the logistic support while working at the field. We acknowledge the available advice and discussion from Professor X. M. Jiang, who showed us his unpublished data. We also give our acknowledgments to those who helped us in the chemical and isotopic analyses.

REFERENCES

1. L. Beccaluva, M. Barbieri, H. Born et al. Fractional crystalization and liquid immiscibility process in the alkaline-carbonatite complex of Juquia (Sao Paulo, Brazil). J. Petrol. 33, Part 6: 1371—1404 (1992)
2. Y. Bottinga. The density of magmatic silicate liquids: a model for calculation. Amer. J.

Sci. 271: 438—475 (1972)

3. L. Corriveau and M. P. Gorton. Coexisting K-rich alkaline and shoshonitic magmatism of arc affinities in the Proterozoic: a reassesment of syenite rocks in the south western Greaville Province. Contri. Mineral. Petrol. 113: 262—272 (1993)

4. M. J. Dorais and C. Floss. An Ion and Electron Microprobe study of the mineralogy of enclaves and host syenites of the Red Hill Complex, New Hampshire, USA. J. Petrol. 33: 1193—1218 (1992)

5. G. Faure and J. C. Powell. Strontium isotope geology. Springer-Verlag Berlin-Heidebery. New York (1972)

6. G. Faure. Principles of isotope geology. John Willey and Sons. 141—247(1986)

7. M. Gao and F. Gao. A study on Rb—Sr isochron age of the granulite in Zhangjiakou—Xuanhua area of Hebei Province. Journal of Tianjin Institute of Geology and Mineral Deposits of Chinese Academy of Geological Sciences. No. 14, 121—127 (in Chinese) (1988)

8. R. B. Hargraves. Physics of magmatic process. Princeton University Press, Princeton, New Jersey (1980)

9. C. M. B. Henderson C. M. B. , K. Pendiebury and K. A. Foland. Mineralogy and petrology of the Red Hill alkaline igneous complex, New Hampshire, USA. J. Petrol. 30: 627 —666 (1989)

10. X. D. Hu, J. N. Zhao and S. B. Li. The vein-gold mineralization in the Archean metamorphic rocks in Zhangjiakou—Xuanhua area. Journal of Tianjin Institute of Geology and Mineral Deposits of Chinese Academy of Geological Sciences. No. 18 (in Chinese) (1990)

11. H. Y. Li, L. Peng and G. F. Wang. The metallogenetic mechanism of the Chongli—Chichen auriferous shear zone, Northwestern Hebei Province. Journal of Precious Metallic Geology, 3: 169—175 (in Chinese) (1987)

12. B. A. Litvinovskiv, A. N. Zanvilevich and I. V. Aschepkov. The nature of the sviatonossits of Lake Baykal. Inter. Geology Rev. 28: 46—61 (1986)

13. B. L. Mu and G. H. Yan. Geochemistry of Triassic alkaline or subalkaline igneous complexes in the Yan-Liao area and their significance. Acta Geologica Sinica. 66: 108—121 (in Chinese) (1992)

14. F. E. Mutschler. The precious metal deposits associated with alkaline rocks——a spatial and temporal process in Cordillera. Mining Engineering. March: 304—309 (1991)

15. J. A. Shao. The crust evolution of the middle part of North Margin of China—Korea plate. Beijing: Peking University Pub. House (in Chinese) (1991)

16. H. R. Shaw. Visicities of magmatic silicate liquids: an empirical mothed of prediction. Amer. J. Sci. 272: 870—893 (1972)

17. R. X. Song. Origin of the Shuiquangou monzonite intrusion and its relationship to the gold mineralization in Zhangjiakou area. Contribution to Gold Deposits (No. 1), 82—91 (in Chinese) (1990)

18. D. J. Tan and J. Q. Ling (eds). Mesozoic potassic magma region of the plateform of North China. Beijing Earthquake Publish House (in Chinese) (1994)

19. H. P. Taylor. The oxygen isotope geochemistry of igneous rocks. Contri. Mineral. Petrol. 19: 1—71 (1968)

20. R. R. Wang. The characteristics and origin of the felsic alkaline complex of the Jinjiazhuang area, Hebei Province. Journal of Guilin College of MetalluryGeology. No. 1: 12 —19 (in Chinese) (1992)

21. Z. K. Wang, X. M. Jiang, Y. Wang et al. Origin and geological significance of the Shuiquangou weakly alkaline complex in Zhangjiakou—Xuanhua area, Hebei Province.

Journal of Precious Geology. No. 1; 14—20 (in Chinese) (1992)

22. J. Y. Wei and J. Su. Geochemical characteristics of the Shuiquangou intrusion in the Dongping gold mine, Hebei Province. Geological Science. 29; 256—266 (in Chinese) (1994)

23. S. Y. Xiang, J. L. Ye and J. Liu. Origin of the alkaline syenite intrusion from Hougou to Shuiquangou and its relationship to gold mineralization. Modern Geology. No. 1; 55 —62 (in Chinese) (1992)

24. Y. C. Xu, Q. Me and S. Z. Li. Study on the gold mineralization in Dongping mine. Geology and Exploration. No. 2; 29—33 (in Chinese) (1994)

25. G. H. Yan, B. L. Mu and Y. S. Zen. The temporal and spatial distribution of the alkaline and weakly alkaline intrusions in North China and tectonic significance. Journal of Institute of Geology and Mineral Deposits of Chinese academy of Geological Sciences. No. 19; 93—100 (in Chinese) (1989)

26. R. E. Zartman and B. R. Doe. Plumbotectonics—— the model. Tectonophysics. 75; 135—162 (1981)

27. G. C. Zen. The petrological and mineralogical characteristics of the Shuiquangou complex of Chongli County, Hebei and its genesis. Geological References of No. 1 Geological Party of Metallurgical Ministry. No. 4, 29—37 (in Chinese) (1992)

28. Z. C. Zhang. Origin of the Shuiquangou complex in northern Hebei Province and a study of its relation to gold mineralization. unpubl. Ph. D. thesis, Graduate school, Chinese Academy of Geological Sciences (in Chinese) (1995)

29. Z. C. Zhang. Discovery of biotite-pyroxene diorite in the Shuiquangou weakly alkaline complex, northern Hebei Province, and geological significance. Acta. Petrol. Miner. 14; 9—18 (in Chinese) (1995)

30. Z. C. Zhang. The characteristics of garnets of the Shuiquangou complex in Northern Hebei Province and geological significance. J. Miner. Petrol. No. 2; 7 — 15 (in Chinese) (1995)

31. Z. C. Zhang and J. W. Mao. Geology and geochemistry of the Dongping gold telluride deposit in North Hebei, China. International Geology Review, 37;1094—1108 (1996)

32. Z. C. Zhang and H. X. Chen. Geology and Geochemistry of the Shuiquangou complex, Northern Hebei Province, China. Scientia geologica Sinica, 5;447—468 (1996)

Milton Keynes UK
Ingram Content Group UK Ltd.
UKHW040059071024
449327UK00019B/659